SCIENCE AND DEMOCRACY

A Science and Technology Studies Approach

Linda Soneryd and Göran Sundqvist

BRISTOL
UNIVERSITY
PRESS

First published in Great Britain in 2023 by

Bristol University Press
University of Bristol
1–9 Old Park Hill
Bristol
BS2 8BB
UK
t: +44 (0)117 374 6645
e: bup-info@bristol.ac.uk

Details of international sales and distribution partners are available at bristoluniversitypress.co.uk

© Bristol University Press 2023

British Library Cataloguing in Publication Data
A catalogue record for this book is available from the British Library

ISBN 978-1-5292-2213-5 hardcover
ISBN 978-1-5292-2214-2 paperback
ISBN 978-1-5292-2216-6 ePub
ISBN 978-1-5292-2215-9 ePdf

The right of Linda Soneryd and Göran Sundqvist to be identified as authors of this work has been asserted by them in accordance with the Copyright, Designs and Patents Act 1988.

Cover design: blu inc
Front cover image: Shutterstock/Omelchenko
Bristol University Press use environmentally responsible print partners.
Printed and bound in Great Britain by CPI Group (UK) Ltd, Croydon, CR0 4YY

FSC
www.fsc.org
MIX
Paper | Supporting
responsible forestry
FSC® C013604

Contents

Preface

This book does two things: it introduces the field of science and technology studies (STS) and discusses science and democracy through an STS approach. This double aim of the book means the book is an introduction to STS with a specific focus.

STS is a lively international and interdisciplinary research field. Despite the heterogeneity of STS as a research field, scholars within STS share some common theoretical grounds. What holds STS together is an understanding of knowledge, especially scientific knowledge, as practice in which actors' interpretations, judgments, and actions lead to results that can be perceived as robust and true knowledge. The authority of knowledge, or truth, is not something that exists from the beginning, but is the result of processes in which many actors participate. It is these processes of emergence that are of interest to all STS researchers.

The research field has several roots, both academic and more social and political. Academically, the scientific theoretical discussion in the wake of Thomas Kuhn's book *The Structure of Scientific Revolutions* (published in 1962, with an important second edition in 1970) has been central. Questions about paradigms, paradigm conflicts and paradigm changes have had great significance outside the world of research as well. These debates led to a return to the classical sociology of knowledge, which was an important part of the emergence of sociology at the turn of the last century and whose tools could now be used with new energy to understand not only religious beliefs, political ideologies and everyday knowledge (folk beliefs) but also scientific knowledge. Scientific knowledge is not dramatically different from other forms of knowledge, and it can be sociologically understood in the same way: as anchored in a particular context, shaped by time and place, and by different interests and actor strategies. But despite these similarities, scientific knowledge can also be very different from other forms of knowledge.

In this book, we take as our starting point the principle that legitimizes science in modern society and makes it different from other knowledge forms, and this principle is also what science and democracy have in common. The principle shared by science and democracy is that their authority is not gained through violence. Rather their authority is based on the principle that they can always be questioned and revised.

Societal changes in the Western world during the late 1960s have also had great significance for how the field of STS research has developed. The social movements that became important, such as the women's movement, the environmental movement, and the civil rights movement, all included a critique of science and technocracy. Science was criticized as a social force that was largely responsible for war (through weapons development), environmental destruction (through chemicals) and oppression (through biological race doctrines and gender systems). The critique pointed out how science gained huge and unjustified influence over issues that should not be understood as expert issues, but rather as political ones.

Theoretically, early STS research was dominated by a clear social constructivism, which later has turned to a more open constructivism, which often tries to avoid sharp divisions between the social and the 'other'. Actor network theory (ANT) is perhaps the prime example of this. Today, ANT is also known and used in academic disciplines outside the STS field.

STS emerged with a strong focus on the question of how scientific knowledge is produced. The important studies from the 1970s were often those carried out in a scientific laboratory environment. In recent times, research in the field has come to be dominated by studies on how scientific knowledge is spread, how it is used and how it affects the world outside the laboratory. And this is where the question of science's relationship to politics and democracy becomes important. The question of governance of society becomes a question of knowledge. The best knowledge is needed for the best management. But who decides what is the best knowledge, and are ideas of governance and knowledge intertwined?

Both of us have been engaged with these issues in our research, and when we became colleagues at the University of Gothenburg in 2015, we decided to write a book together. That decision resulted in two books in Swedish: one monograph on science and democracy (Sundqvist and Soneryd 2019, *Vetenskap och Demokrati*, Studentlitteratur) and one edited volume on scientific citizenship (Soneryd and Sundqvist 2019, *Vetenskapligt medborgarskap*, Studentlitteratur). The book *Science and Democracy. A Science and Technology Studies Approach* is a revision and compilation of those two books. This book has the aim of introducing both classical STS work and more recent discussions around post-truth, for example.

Our own research has focused on the science–policy interface and on public participation. Many of the empirical examples in the book are from our own research in areas such as nuclear waste siting, climate change, controversial technologies and other environment-related issues.

The wide topic of science and democracy is of interest to anyone who is concerned with today's global challenges. Science, and not least technological development in symbiosis with science (the so-called technoscience that includes everything from biotechnology to artificial intelligence and geoengineering), is becoming an increasingly important social force. The main focus of the book is governance of and controversies surrounding science and technological developments. In this context, science usually means natural sciences and medicine (rarely social sciences or humanities, although social science expertise can be crucial for governance). It is in relation to this topic that we discuss democratic shortcomings and opportunities for democratization. This means that there are a number of important issues related to democracy that are not primarily related to science and are therefore not touched on in the book.

Democracy can never be taken for granted, and democratic values can be threatened by the development of technoscience but also by authoritarian and populist forces. But the direction of science and technology developments can never be predetermined. The relationship between science and democracy takes a number of different forms of expression, and these will continue to develop in the future. And as those involved in STS research often put it, science and democracy are practices that are made, and in this making, we are all co-creators. Our hope is that this book can contribute to an improved understanding of how both science and democracy are made in practice.

We wish to express our thanks to a number of colleagues who have read drafts of the manuscript and helped to improve it: First, we wish to express our thanks to Paul Stevens, our editor at Bristol University Press, who took on this project with great enthusiasm. We also owe thanks to three anonymous reviewers who gave us crucial input on our initial outline, and later on also a review on our full draft of the book gave us constructive feedback. Our colleague Abby Peterson read the English manuscript in full. Mikael Carleheden and Catharina Landström read and commented on the Swedish version, and this input guided us when compiling and revising the English version. Rune Premfors has long been a discussion partner on issues concerning democratic theory, and he also gave some input to this book, though he may not be aware of it. The master's course on Scientific citizenship that Linda Soneryd was invited to give at the STS department, University of Vienna and the discussions this enabled with the students, Ulrike Felt, and Alan Irwin, was fundamental for how Chapter 7 developed. We are also grateful for our previous collaboration around the edited volume that we compiled into Chapter 7, and especially thanks to Andreas Gunnarsson, Jesper Petersson and Johan Söderberg for giving us the permission to draw on their work to provide examples. We also want to thank Kristin Asdal, Yannick Barthe, Erlend Hermansen and Rolf Lidskog for valuable inspiration and good cooperation. Parts of the manuscript have been presented at the Score (Stockholm Centre for Organizational Research) seminar, at the sociology seminar at Örebro University and at the conference STS Sweden. We are thankful for all constructive comments and discussions that have made working on this book a pleasant journey.

Linda Soneryd, Örebro University
Göran Sundqvist, University of Gothenburg

1

The Best Knowledge and the Best Mode of Governance

Introduction

Science, in its very basic sense, means knowledge, from the Latin *scientia*; and democracy, in its very basic sense, means rule of the people, from the Greek *demos* (people) and *kratos* (rule). Throughout history, science and democracy have each developed with a keen ability to alter their shape in various contexts. Nevertheless, science has come to denote specific ways of producing knowledge as opposed to others, and democracy has come to be associated with particular modes of governance as opposed to others. The reason why it is impossible not to use the plural in relation to the practice of these two concepts – *ways* of producing knowledge and *modes* of governance – is simply that there are multiple versions and ways of doing both science and democracy. Yet science and democracy have taken shape as *the* grandeurs of modern societies. This suggests that there is something fixed about them: science has become established as the best way of producing knowledge, and democracy has become established as the best mode of governance.

Of course, their grandeur refers to the general ideas and principles that are associated with both science and democracy. Even though these ideas and principles have generated deep thought, lengthy discussion, and many thick books, dominating versions of both science and democracy can be summed up in one word: *representation*. Scientists represent nature, and elected politicians represent the people. To gain legitimate authority, however, representations and representatives must resonate with the represented. It is this resonance and the relational aspect involved in both science and democracy that distinguish them as particular ways of producing knowledge and particular modes of governance. The principle behind both science and democracy is that authority is not gained through violence and power exercises that ignore objections to the representations if they fail to be 'fair', 'just', 'correct', 'reasonable', 'relevant', and so on. It is this principle – that the represented can *object* to representations and that their objections (*if* assessed as legitimate) will induce relevant changes – that replaces gods (or emperors) and nature as unquestionable authorities, and it is therefore science can provide us with the best knowledge and democracy with the best mode of governance.

Representation, as a term and as an idea, is filled with all sorts of associations and predefined meanings. These preconceptions also imply that representation means something different in the context of science compared to what it means in the context of democracy. Representation is central in this book, but our take differs from many established understandings of its meaning. Our approach to representation is neither to be found in the philosophy of science, nor does it give justice to all the nuances that we can find in political philosophy and debates around democratic representation. Rather we *sociologize* the question of representation and see it as a particular social practice and a continuous and changing relation between representatives and the represented, between governing elites (of scientific experts or politicians) and concerned publics. Yet, we also admit that the relations within the practices of science and the practices of democracy are different; while natural objects constantly object to the scientists' claims (by not behaving as expected), human beings can be manipulated and disciplined and hence rendered less able to object (Latour 1999a: 10).

In this book, we present, but also problematize, ideas about science and democracy and the relation between them. Intellectual interest in the interplay between science and democracy has a long tradition, but we primarily treat this relation from a relatively young academic tradition: science and technology studies (STS) (for overviews, see Asdal et al 2007; Bucchi 2004; Felt et al 2017; Felt and Irwin 2023). The questions around knowing and governing that STS analyses in its own specific ways have for a long time been central in academic disciplines such as sociology, political science, philosophy, and social theory in general. With this book we would like to discuss, more specifically, what STS can contribute, using analyses of the interplay between science and democracy.

It should be said that this book is not as much about democracy as it is about science. Science sometimes plays a crucial role in areas of public concern, but at other times the role of science remains more marginal. Three central structural problems that all democracies need to deal with are social inequalities, minorities whose identities and life conditions are not recognized, and the political complexity that is implied by increased specialization and expert dependencies (for example, Fraser and Honneth 2003; Carleheden 2009; Fischer 2009). We are primarily occupied with the latter theme, which has to do with how knowing is intertwined with the political. This means that our discussion of democracy is limited, given that science is the central focus for STS research.

The book takes the reader on a journey that begins with the differences and demarcations often set up between science and democracy, and more generally between knowledge and politics, and then step by step introduces the similarities and interdependencies that can also be found. The interplay between science and democracy is complex – characterized by both similarities and differences, closeness and distance – and our ambition is to sort out how these complex interactions can be understood from an STS approach.

Our journey in this book begins in Part I, titled Separation, continues in Part II, Overlap, and ends in Part III, Co-production. Before this journey starts, in this first chapter we provide a background to the discussions that follow. As already mentioned, we focus on what science and democracy have in common: they are both about *representation*. We discuss how representation is understood in relation to first science and then democracy. Even if representation is important for both, we often intuitively think about science and democracy as working according to opposite logics. Representation is interpreted differently in science compared to democracy. Science should not be democratized, and democracy should not be scientized. Science is elitist and based on demarcations between those who are scientific experts and those who are not. Democracy, in contrast, is egalitarian and based on the idea that we are all equal. In the section that follows, we counteract these intuitive notions and begin to unravel what science and democracy have in common. From the background of increased *political complexity*, we discuss *post-truth* and *participation* as two themes that potentially problematize the relation between science and democracy. We also link these themes to current discussions within STS.

Science and democracy as representation

Science and democracy have common historical origins. During the 16th and 17th centuries, scientific discoveries made such an impact on our ways of thinking about the world that this period has come to be known as the time of 'the Scientific Revolution' (Shapin 1996). Since then, scientific practice has been guided by principles of observation, experiment, and criticism, which later on were reformulated into methodological rules. A systematized scientific method was seen as key to the production of reliable and shared knowledge. Similarities between how science and democracy have developed can be seen in the struggles against sovereign authorities claiming to represent gods or nature from a position that was assumed as given.

The development of modern democracies is, therefore, difficult to understand – and perhaps impossible to achieve – without relating it to the development of scientific reason. The common roots are particularly clear in the 18th-century Enlightenment project, through which new ideas about the origins of knowledge developed in parallel with the political revolutions in England (1688–1689), the United States (US) (1775–1783), and France (the Revolution of 1789) and the establishment of national parliaments that limited, or abolished, the power of the kingdom.

The ideals that emerge through these historical events can be summed up through the idea of *representation*. Science represents nature, and democratic governance represents the will of the people. The scientific method was assumed to secure the objective character of the natural sciences and enable scientists to set aside their passions and interests in the making of scientific knowledge. Together with publicly elected parliaments, modern nation states are based on the separation of powers (for example, executive, legislative, and judicial powers) to ensure that the arbitrariness of rulers can be avoided. The authority of scientific knowledge and the legitimacy of democratic governance are, however, both dependent on continuous recognition by others and, hence, recognition of potentially alternative representations.

Science and democracy thus share important principles and ideals that oppose illegitimate claims for knowledge and power (Ezrahi 1990, 2012). The authority of both scientific knowledge and democratic governance is legitimated by representation. It is only when this representation is accepted as legitimate that it can support the authority of the knowledge claims made, or the authority of particular forms of governance. Representation is, however, never perfect or complete. Since both scientific practices and democratic governance, *in principle*, admit potential alternative representations, continuous struggles over what is represented and what is not can be foreseen. To represent means to simplify, which implies the existence of alternatives – alternative ways of simplifying complicated issues. This further means that both science and democracy have inbuilt mechanisms for improvement, which are based on critical assessments of existing representations. Next, we discuss the idea of representation in more detail, as always in principle being possible to reassess and improve, focusing on science and democracy in turn.

Science

Steven Shapin and Simon Schaffer (1985) describe the development of modern science in a society in transition. Their prime example is the emergence during the 17th century of the scientific experiment as a crucial part of scientific practice and how this generated scientific conflicts at the same time as political–democratic conflicts were taking place in relation to the restoration of the monarchy in England. Experimental science and its reliance on observation provides opportunities to criticize the established dogma and epistemic power of the scientific elite. Consequently, what is said by scientific elites is not always true, and it can be counteracted with empirical evidence (Ezrahi 1990). Witnesses become central, as seeing is more important than believing.

In other words, scientific representation can always be questioned and potentially altered. The fact that representation can be changed, and is based on observations instead of established beliefs, makes it seem both equal and democratic, but this is complicated by the fact that the scientific elite has a privileged position also in relation to having the ability to propose what is observed. This interpretive prerogative and the fact that not everyone can make themselves heard may lead to some being incapable of claiming the right to be represented. To observe is thus never viewing from nowhere, but conditioned and relative to a particular position, for example, by access to observational instruments and schemes of interpretation. Therefore, elites are in a good position to claim that they are the ones who can correctly understand – that is, observe, interpret, and represent – the outcome of an experiment.

That science produces representations of objects that exist independently of these representations is a classical idea in Western societies since the Scientific Revolution and is also part of common-sense knowledge. This idea includes a realist position about an external world. Knowing this world, however, is dependent on observations and interpretations. Consequently, debates can go on about accurate and false interpretations of phenomena even though they are assumed to exist independently of researchers and their methods and instruments. Scientific practices are full of these quarrels, and this is a prime topic in the philosophy of science, where we find a lot of different positions regarding basic agreement about scientific knowledge being about representing objects in the natural world.

For our purpose in this book so far, it is important to emphasize that this classical idea of representation in science is based on the idea that the world could be represented differently. However, we should also note that there are constructivist social scientists who oppose this representationalism. They argue that the objects that scientific knowledge are 'about' are constructed or enacted, but *not* represented. The 'external world' is not independent of, but dependent on, the research process itself. The idea that the world that science is about is constructed, rather than represented, is a position that can be termed 'anti-representationalism' (Hammersley 2022).

As stated earlier, our approach to representation is not found in the philosophy of science; nor are we interested in criticizing this position from an anti-representationalist point of view. Our take is different, and here we follow a classical position in STS which means keeping the interest in representation but bracketing the question of what is correct and true in relation to the external world. This STS position implies a division of labour between scientists, who try to produce knowledge representing nature (natural objects) by making interpretations, and STS scholars, who study the interpretations made by the scientists from their social conditions. The latter position means that we understand representations as results of social processes without denying that they are also connected to processes in the natural world. This position assumes that there is *also* a connection between representations and social conditions, which makes this relation an important object of study. This is the classical STS position formulated in the 1970s by David Bloor (1991) in the 'strong programme' of sociology of scientific knowledge.

When scientific representations are understood as socially conditioned, this implies that conditions could be otherwise and that, consequently, the interpretations could be made differently. Thereby, these conditions explain why scientists interpret differently. Social conditions vary, as do the ways human beings are socialized into specific social groups and cultures, such as scientific communities and research groups. Representations always need resonance and acceptance from social groups. In fact, agreed knowledge (representations) always includes group formation, as STS scholar Harry Collins (1992: 159) specifies in relation to how social conditions could explain how individual interpretations become scientific facts.

However, the self-understanding of modern science supported by the philosophy of science is that problems of representation can be overcome. This assumes that once the knowledge in question has been critically reviewed by colleagues and thus approved, the representation has also been approved as correct. This idea of an approved end point of knowledge has, throughout history, been expressed in a variety of ways: as objective or verified knowledge or as the establishment of a universally valid truth.

We partially share this standard view of science: scientific knowledge is about representation and the scientific method needs to be transparent about how observations and interpretations are made. However, we also problematize scientific representations by directing our attention to conflicts and struggles over representations – that is, how representations are made, supported, criticized, and maintained – which sometimes implicate the existence of authoritative (agreed) representations. Struggles over scientific representations are indicative of scientific practice, and there are no end points to arrive at.

However, these everyday struggles could manifest as controversies. Among these, the most common type that we come across in news reporting, in public debates, and in our own research is the *mixed controversy*. A mixed controversy is never solely about scientific representation, but is also concerned with issues related to societal development, responsibilities, inequalities, and justice.

Examples of a mixed controversy can be found everywhere, and as an illustration we can think about the COVID-19 pandemic that broke out in 2020.

Soon after medical scientists discovered the coronavirus, pictures of the virus frequently accompanied news headlines. A considerable number of resources and activities were directed to producing a vaccine. When the pandemic had ravaged for about nine months, a new vaccine was presented, which was scientifically assessed as effective and safe and then approved and subsidized by governments.

Before the vaccine was rolled out, many other measures were taken by governments and responsible actors, such as recommendations to keep socially distanced, wash hands, and wear masks, obligations to report illness to the authorities, and not least lockdowns of most activities in society. These activities were responses to manage the challenges of overloaded healthcare systems. Debates took place around all these measures, and much critique was directed to governments and how they managed the pandemic. A central element in these debates was whether the measures were motivated by science *or* were demonstrations of political power, and not least what a good mix of scientific and political considerations should look like.

A common and typical way of supporting and reproducing a separation between science and society is to maintain a narrow scientific framing even when critical voices attempt to present the issue as multifaceted and complex. This means trying to prevent a mixed controversy from emerging. For example, a narrow framing can be maintained by referring to scientific evidence for a particular approach or by supporting the argument through reference to expert knowledge, even when the argument goes far beyond that specific expertise. Yet, to protect the pre-eminence of science in such situation is not easy. When a broader set of actors and issues have appeared on the agenda – which happens when scientific results leave the laboratory – the scientific issue is modified and becomes mixed and can no longer be answered by experts or technical solutions alone. In public discussions on how to respond to the COVID-19 pandemic, we have seen many examples of different framings in practice, more or less scientifically motivated (Horton 2020). Such mixed discussions (or controversies) can lead to a democratization of processes that are highly dependent on expert knowledge.

We argue in this book that science is not separate from society. What is deemed science can be jeopardized. We can see this in the existence of gender studies, studies of ethnic minorities and LGBTQ+ research being threatened by authoritarian regimes. However, a postulated a priori conflict between science and society is not supported by such examples. Instead, science is a fundamental part of society, continuously adding new ingredients to it. Hence, scientists do not typically solve complexity, but rather add to complexity. According to STS scholar Bruno Latour (1998: 209):

> No one, for instance, believes that ecological controversies will die away to a point where we will no longer have to take care of the

environment. Activists as well as scientists and politicians do not expect science to decrease the complex web of their lives. On the contrary, they expect research to multiply the number of entities with which they have to deal in their collective life.

This means that science is an increasingly important part of *society in the making*. In the next section, we build on this when presenting our take on democracy as representation.

Democracy

As already explained, in this book we are not focusing on 'philosophy of science' debates about scientific representation. Similarly, our primary interest is not political philosophy and debates around democratic representation. This has already been worked on in depth by others, such as Mark Brown (2009), who, in the book *Science in Democracy*, exhaustively discusses the philosophical roots of a range of different ideas about democratic representation and in relation to scientific practices.

Instead, we start with a simple question and two equally problematic answers: What is democracy? The first answer is that democracy is the rule of the people. The second answer is that democracy is a complex of institutions, including representative government, that, taken together, is referred to as 'democracy'. Both answers are problematic since what 'the rule of the people' means is today very different from the time when democracy, as a term and a as mode of governing, was first invented in ancient Greece. To speak of democracy as a set of institutions can also be problematic, since these institutions do not form a coherent whole. The argument that democracy is an amalgam of elements that do not fully cohere is made by Robert Dahl (2008) in his famous book *Democracy and Its Critics*. In this, he describes democracy, and its theories and practices, as originating from many sources. The four most important sources are: the classical Greek democratic city-state; the republican tradition derived from Rome and the Italian city-states of the Middle Ages and the Renaissance; the idea and corresponding institutions of representative government; and the logic of political equality (Dahl 2008: 13). The result of this is 'a jumble of theory and practices that are often deeply inconsistent. What is more, a close look at democratic ideas and practices is bound to reveal a considerable number of problems for which no definitive solution seems to exist' (Dahl 2008: 2).

It is Dahl's conviction that an analysis of the possibilities and limitations of democracy only can be made in light of the criticisms of democracy – that is, to take seriously the unsolvable problems that the theory and practice of democracy reveal. Dahl thereby commits to a view of democracy as always imperfect, an unfinished project, or, in other words, a process. In addition, Dahl argues that the 'half-hidden premises, unexplored assumptions, and unacknowledged antecedents' of democratic theory 'form a vaguely perceived shadow theory' that accompanies democratic theory wherever it goes (Dahl 2008: 3).

In a similar vein, more recent thinker Hélène Landemore (2020), in her book *Open Democracy*, describes the problems of democracy as internal. She argues that the most dominant idea of democracy – representative democracy – has privileged the *consent* of the public to be represented, as opposed to putting an emphasis on the *power* of the people. Landemore therefore develops an idea of open democracy which is based on a reaffirmation of the value and meaning of people's power. Open democracy is based on the equal chances of each citizen to become a representative, and it emphasizes citizens in the capacity of being simultaneously ruler and ruled, representing and being represented.

In the context of democracy, our take on representation is about an ongoing and continuous relation between governing bodies and concerned publics, which does not limit governing bodies to elected politicians, but potentially includes other actors, such as industry, expert organizations, and advocacy groups. In emphasizing the relational aspects, we also highlight the power of the people to withdraw their consent and to question the actions or decisions made by governing elites. However, one important difference between our approach and that of thinkers in the field of democratic theory is that we are not evaluating existing democratic institutions, nor are we suggesting any new mechanisms for reinventing and reinvigorating democracy, such as new systems for voting or establishing representatives.

In the light of Dahl's famous analysis of democracy and its critics, our approach means to interrogate, using empirical case studies, some of the half-hidden premises of democratic theory. We, like many other STS scholars, have carried out critical analyses of who or what governs expert-based issues, as well as how publics are constructed by the very form of governing. STS scholars have pointed to internal problems with democracy as a system of representation based on a small ruling elite and a large group of citizens excluded from power.

As examples of studies in this vein, we find STS studies of how publics can emerge in relation to highly complex scientific issues (Marres 2007); how technical choices are part of processes of creating and stabilizing public problems, and hence at the core of issues that, ideally, are democratic issues (Laurent 2017); how issues enter parliaments (Asdal and Hobæk 2016); and how the materiality of parliamentary politics matters for how political realities take shape (Danyi 2018) and critically engage in the machineries of representation in the voting system (Lynch et al 2005) and in the imaginaries of how both humans and non-humans can be more democratically represented (Latour 2004). Our book focuses on certain sites of democracy, and in particular we explore various sites of science–policy interactions. These relations frame what issues appear on a political agenda or are subject to public debate, and determine who can be involved and on what terms.

Admittedly, we are living in times when citizens have 'lost much faith in once precious political and rhetorical resources such as the master narrative of progress' and 'the power of science to define reality and induce consensus' (Ezrahi 2012: 301). Because of this, it is of great importance to engage with new political imaginaries of public reason and democratic rule, and our focus on how issues are framed in science–policy interactions, including mixed controversies, is a way of doing this. We believe that STS scholars have important insights to add to such efforts due to the political complexities that have arisen because of ongoing developments in science and technology.

Political complexity and technoscience

Today's pace of technological change and high specialization poses new challenges to democracy. For example, artificial intelligence (AI) as support for decision-making raises ethical and moral questions about responsibilities. Managing global challenges – such as climate change, the loss of biological diversity, and continued increase in the use of chemicals – require scientific expertise, but these challenges have no simple solutions, and it is not obvious what kind of expertise can support more sustainable development. Pressure to adopt certain solutions – such as emerging technologies for carbon capture and storage, electrification, or the adoption of AI in public services – on a policy level or through reprioritized investments may be strong even when the solutions are not accompanied with a clear idea of the implications of their adoption (and sometimes not even a clear idea of when and how they can be implemented). As a result, expert organizations and commercial knowledge industries can have a huge impact on what happens even without having a clear political mandate or there being critical discussions among concerned publics at the outset.

In the introduction to this chapter, we presented the political complexity that follows from increased specialization and expert dependency as one of the structural problems of democracy. Science and technology developments can make it harder to identify who or what is governing, which also contributes to new inequalities and power imbalances as well as making it more difficult to demarcate who is concerned or affected. This complexity is further strengthened since technology developments, as well as associated impacts, escape the politico–administrative boundaries of the nation state. These issues correspond directly to what has been discussed as the vulnerabilities of democracy (Gustavsson 2018) – which are about difficulties in identifying who rules and problems with demarcating concerned publics – and, as a result, increasing social inequalities. As Dahl (2008) wrote, it is only through continued discussion and political development that we can prevent these vulnerabilities from being ignored and leading eventually to democratic breakdown. Scientific modes of knowing are intertwined with modes of governing. When issues are scientifically framed and highly expert dependent, this motivates a special focus on how issues pertaining to science and technology can be subject to democratic control. This is a far more complex process than is captured by the view that 'science' and 'democracy' are separate entities and even opposites. We find the relationship between 'science' and 'politics' taken to extremes in debates around Donald Trump and Greta Thunberg.

From complexity to over-simplification

In October 2017, when Donald Trump had been US President for 263 days, *The Washington Post* (2017) reported that during this time the president had made no less than 1,318 false or incorrect statements. The statements ranged from being false and relatively simply refuted with reference to reliable sources, to being exaggerations to make his own political moves appear nothing other than excellent. In addition, President Trump also questioned established science. He called global warming fake, he terminated US participation in the climate agreement decided in Paris in 2015, and he appointed an explicit climate denier as administrator of the US Environmental Protection Agency.

In 2016, the same year as Trump's election campaign was going on, there was a referendum in the UK about ending its membership of the European Union (EU), known as Brexit. The information campaign that preceded the referendum was accused of being full of exaggerations and untruths. In the aftermath of both the US presidential election in 2016 and the decision about the UK's exit from the EU, terms such as 'post-truth', 'alternative facts', and 'knowledge resistance' have gained pervasiveness (Klintman 2019). With false statements presented in public even though they can be easily counteracted by reliable sources, it seems that we have moved a long way from the Enlightenment ideal of reason. Terms like 'post-truth' and 'fake news' have started to flourish, and the US presidency under Donald Trump became the epitome of a 'relaxed' relationship with the truth.

One explanation put forward for this emerging post-truth era is the increasing influence of critical science studies, that is, studies that go behind the official presentations of scientific results and show how these results are produced and should be understood as conditional to historical, social, and cultural circumstances (for a nuanced discussion about STS and post-truth, see Lynch 2017). In other words, there are those who claim that efforts to open up the black boxes of scientific practice for public scrutiny – something that we, the authors to this book, subscribe to as an important academic ambition of enlightenment as well as an important element of a democratic society – have actually caused the emergence of a post-truth era based on the relativization of truth.

In contrast, the need to take science seriously is a repeated message in the global mobilization for stronger attention to and political action to combat climate change. Now a global movement, Fridays for Future grew out of a school strike for the climate outside the Swedish parliament in Stockholm on 20 August 2018, organized by 15-year-old Greta Thunberg. Similar school strikes then spread all over the world. In 2019, Greta Thunberg was named Person of the Year by *Time* magazine and one of the ten most influential people in science by the journal *Nature*. In September of that year, she spoke in front of the US Congress. Basing her testimony on the Intergovernmental Panel on Climate Change special report on the possibilities of reaching the 1.5 degree target, she expressed the words: 'I don't want you to listen to me, I want you to listen to the scientists' (*The Guardian* 2019).

Are we presented with two opposite stories here, personified by Donald Trump and Greta Thunberg, or are these two tendencies somehow part of the same story? How can we make sense of these stories when thinking about the relationship between science and democracy? There are those that argue the post-truth era is a case that proves that democracy is fundamentally flawed and necessitates a rule by science (Stevenson and Dryzek 2014). In a post-truth era, the authority of science is threatened, but so is democracy, or at least a particular view of democracy that rests on the idea of public reason, and in the examples of President Trump and Greta Thunberg we find complexities in both science and democracy reduced to what could be called over-simplifications. In the following sections, we discuss 'post-truth' and 'participation' as two much-noted – not to say controversial – themes that are linked to the relation between science and democracy as well as to current discussions within STS on how they should be understood. These sections also aim to clarify the normative positions we take in this book in relation to science and democracy.

Post-truth and STS

In studies of knowledge production in action, STS researchers show that truths are social activities and, like all other social activities, are based on judgements, emotions, and interests. Truths are created by people (researchers), and truths will change over time. Even truths must be backed up with arguments and not by reference to higher powers that are taken for granted, such as gods or nature. The production of truths takes place at particular sites and from particular viewpoints, from which it is not possible to see all relevant facts or take every possible circumstance into consideration. STS researchers have discussed the consequences of this sociological understanding of scientific knowledge and the implications of advocating more democratized knowledge production. Is it this sociological understanding of science and democracy, based on uncertain and questionable – that is, provisional – representations, that undermines the privileged position of science and – at least implicitly – supports a post-truth society?

Even without social studies of science, however, science would occasionally reveal itself as a social activity. Hence, we do not need constructivist studies of science to know that science is conducted by people driven by private interests and even corruption at times. This is an obvious side effect of scientific practice that is both visible and acknowledged during scientific scandals related to fake scientific data or reporting. But these events are relatively rare, often assessed as exceptions that prove the rule of disinterested science. More important than the few research scandals and frauds, however, is the acknowledgement of science as only representing parts of reality. Since science is only conducted from a specific site and in a specific context, it can only represent what it is possible to observe from this position (Haraway 1996). In addition, science has, with its far-reaching specialization, an in-built blindness. This dimension, which is of great importance for STS research and its focus on sites, practices, and contexts of knowledge production – science as simplified representations, as we talked about earlier – is however seldom discussed when this research is accused of opening up the gates to post-truths (Sismondo 2017).

Harry Collins and colleagues point to symmetrical analysis – an important methodological approach in STS that puts brackets around whether or not the knowledge being studied is true and instead focuses on the conditions (including political factors) important for producing the knowledge – as succeeding in making the social character of science visible. However, they also argue that a consequence of this is detrimental. By revealing the continuities between science and politics, STS has 'opened up the cognitive terrain to those concerned to enhance the impact of democratic politics on science but, in so doing, it opened that terrain for all forms of politics, including populism and that of the radical right wing' (Collins et al 2017: 581).

This argument can be summarized as follows: defending a sociological understanding of science may have unwanted effects. But against this argument, one could claim that defending a view of science as objective and universal also has unwanted effects – for example, blindness to gender differences in medical research could lead to presenting conclusions from a study as valid for the entire population even though they are based on data only from men in the population. Can it really be the case that admitting to the social dimensions of knowledge production inevitably opens up the nightmare of populism and the view that all beliefs are equally true? If, due to the possible negative impacts, social scientists were to be prevented from studying how interpretations and interests matter for scientific knowledge production, then science would clearly be subordinated to politics.

Since Thomas Kuhn's 1962 book, *The Structure of Scientific Revolutions*, social science research about science has assumed that the knowledge we hold to be true today is based on assumptions that are not openly recognized and problematized, but rather taken for granted. Kuhn (1962) argues that scientific knowledge is based on a scientific doxa, a so-called paradigm, which is shifting, and sometimes also in a revolutionary way. These paradigms, constituting practices within scientific disciplines, are never free from social interests and other historical and cultural conditions, which science cannot be liberated from and that STS research makes into an important object of study. Rather than denying the social conditions of science, these should be admitted and their consequences for knowledge production should be openly discussed and critically analysed (Bloor 1991; Sismondo 2010: chapter 2).

Despite the tendencies towards seeing expert knowledge as broadly distributed – with scientific results disseminated, critically discussed, and used in society – and the increase of knowledge producers and phenomena that can be reported as post-truth and fake news, science is still a privileged institution with high epistemic authority. The debates around post-truth provide evidence. Fears raised in connection to debates about post-truth and alternative facts are often combined with a support of the traditional image of science as free from political involvement. Perhaps these reactions can be explained by 'fear of mob rule', which assumes that 'reality depends on whatever the mob thinks is right at any given time' (Latour 1999b: 7). The challenge that STS researchers face is thus to provide an accurate description of how science works – as a social activity carried out by people who make assumptions that are not problematized, make assessments that are based on values, and are exposed to social influence and political and economic control – while at the same time sticking to the idea that there are commonly accepted truths and knowledge that are credible and which we have good reason to hold as truths until proven otherwise (cf Shapin 2010). We aim, throughout this book, to develop such a perspective.

Participation and STS

The idea that all those affected by a decision should be able to influence it has been a concern for many STS researchers, albeit one that is not always explicitly articulated. Conflicts around planned buildings or infrastructures, diverging perspectives on the benefits or risks of emerging technologies, and concerned groups that raise issues around effects on human health and the environment – all these examples are permeated with a strong component of specialized expert knowledge. STS analyses of such processes have often focused on contestations of narrow expert framings, and not seldom engaged in giving unheard perspectives voice. For this reason, STS has been described as favouring the normativity of participatory democracy at the expense of other concerns (Pallett and Chilvers 2022), although sometimes STS is described as being in a void of explicit political–theoretical commitments (Blok 2022). Alfred Moore claims that the STS argument in favour of participation can be summarized in the following formula: 'the technical is political, the political should be democratic, and the democratic should be participatory' (2010: 793). In this allegedly naive approach to participation, more voices are always better than fewer voices.

In this book, rather than defending a naive approach to participation or suggesting that we are open to all possible variants of democracy, we would like to make explicit the normativity that comes with the specific view on knowledge that we present in this book. We argue that while this normativity does *not* specify the approach one should take to democracy, it does *align better* with some conceptions of democracy than others. We suggest that a conception of democracy that emphasizes deliberative qualities is a good ally for STS scholars who are interested in how issues and publics emerge in relation to each other – for example, in environmental conflicts. Similarly, Robyn Eckersley describes the reasons for ecological democrats' preference for deliberative democracy, arguing that this democracy

> is not contained by fixed borders, enables communication across the expert/lay divide, welcomes different kinds of knowledge, facilitates social learning, and promotes generalised interests by weeding out purely self-serving arguments through the requirement of answering to others and providing reasons that can be accepted by differently situated interlocutors. (Eckersley 2020: 220)

Another way of highlighting a possible affinity between the deliberative qualities of democracy and STS is that neither the authority of scientific knowledge nor the legitimacy of democratic governance are set once and for all, but are dependent on continuous critical assessment, objection to misrepresentations, and recognition of legitimate claims for corrections. This means that both science and democracy need to be understood as ongoing and unfinished processes and as continuous social struggles. These thoughts are important in approaches to deliberative democracy as well as for the way knowledge production is understood in STS.

In line with this, we rely on the definition of democratic representation as a continuous process of constructing a relationship between governing bodies and concerned publics, involving expectations about citizens' rights and duties as well as imaginations about the public good and good governance. In democratic theory, this definition corresponds to a communitarian tradition that puts more emphasis on the relational aspects of representation – as opposed to a liberal tradition that puts more emphasis on the formal aspects of representation – and the capacity to represent someone as acting for someone and on behalf of someone. Being STS scholars, we are interested in instances where the relation between citizens and government is activated and in issues in which science and technology are important ingredients. We focus on two key entry points to such instances, both of which are classical approaches in STS research.

One entry point is to study how governing bodies discern and relate to relevant publics in matters where science and technology are involved. Here, we find examples that range from clear separation between experts and citizens to varying degrees of interaction between them. Nuclear power is a typical example where a strong separation between experts and citizens might be identified. Ian Welsh and Brian Wynne (2013) illustrate this with a picture from the 1950s of a nuclear power station surrounded by a halo. There are no people, no citizens in this image. But this does not mean that the attitudes of the population were irrelevant. On the contrary, it was decisive that citizens too were enchanted by the promises of nuclear power, but the public had to accept that this issue was delegated to experts. This kind of passive citizenship illustrated by silent acceptance may change if citizens do not continue to accept the delegation to experts or in other ways disapprove of government performance, which has happened many times in relation to nuclear power.

Because of the vulnerability of the relationship between governments and their citizens, governments must listen to publics and find ways to interpret their acceptance or non-acceptance. In this endeavour, governments may, to various degrees, engage in so-called active listening in government-led citizen dialogues, focus groups, or surveys (Lezaun and Soneryd 2007; Metzger et al 2017). Such initiatives for active listening may in themselves be questioned for being narrowly framed, non-representative of public opinion, and excluding important issues and publics. Such issues have been an important topic for STS scholars.

The other, and related, entry point is when citizens engage in expert-based issues as a result of their discontent towards government performance or discovery of problems hitherto unrecognized by governing bodies. STS literature is full of examples, often concerning environmental risks, emerging technologies, and medical health. Noortje Marres (2005, 2007) conceptualizes this by focusing on the intertwined duality of issues and publics. While Marres explicitly speaks to democratic theory, she turns against the established view that democracy deficits appear when politics is somehow displaced and escapes democratic rule, as if politics could ever be contained in democratic arrangements. Instead, with the help of Walter Lippman and John Dewey, she develops an approach to democracy as a relational affair between governing bodies and publics. Publics emerge together with issues; it is when things do not work – when uncertainties, risks, or negative outcomes affect citizens' and communities' everyday lives – that people mobilize as publics. Publics mobilize due to their engagement with issues – that is, the matters that concern them directly or indirectly.

Democracy is about representation – it consists of formally elected and constitutionalized representatives and represented, as well as a continuous process of imaginatively constructing a relationship between citizens and their representatives, between publics and governing bodies, between those who are concerned by decisions and those who make them. Our take on democracy as representation is to follow the classical STS approach of pursuing critical studies of instances and cases where the relation between citizens and government is activated and in which science and technology are important ingredients. We broadly define government as distributed power – and thus not limited to elected politicians – and something that includes a focus on relevant governing bodies, such as expert organizations and industry actors.

To summarize, in this book the classical view of scientific representation and the classical formal view of political representation are understood as being of a sociological character. Consequently, we focus on the relational, contextual, and social organizational aspects of how science comes to represent factual issues, and how governing bodies relate and respond to their citizenry and vice versa. In line with this, we focus on the social conditions of both science and democracy. This strengthens the view that science and democracy have much in common. Nevertheless, how we understand the relation between science and democracy is changing, as advanced knowledge production and expertise is contributing to heightened political complexity.

The structure of the book

This book explores the relationship between science and democracy from an STS perspective. STS is a field which we both have contributed to with our research on science–policy interactions and public participation in the area of controversial technologies and environmental governance. Through the focus on the interplay between science and politics and the role of participation when it comes to highly expert-dependent issues, we contribute important insights into the relation between science and democracy. With this book, we also introduce STS as a field of research, and we present an overview of classical and current STS scholarship and debates from our special focus on science and democracy.

Our main argument is that science and democracy belong together – they are co-produced – *and* are separate. This argument, however, is based on neither a wish to marginalize science, as in the example of President Trump's relativization of truth, nor a wish to prioritize science and enthrone it as a direct guide to political action. Our ambition is more modest; it is to describe how the interplay between science and politics actually works and how this interaction can be better understood through STS analyses, not least the fact that is characterized by diversity and alternative forms of expression.

The book is structured in a way that takes the reader on a journey that starts by emphasizing the differences between science and democracy and then – based on the argument in this introductory chapter – piece by piece, introduces their similarities and interdependencies. Therefore, this book takes us on a journey through three parts: Separation, Overlap, and Co-production.

Part I, Separation (chapters 2 and 3), begins with the canonical idea in Western societies that science and democracy work according to different logics. Chapter 2 deepens this notion by turning to ancient Greece and Plato's view on the matter. We also describe how sociologists have thematized premodern, modern, and late modern societies according to increasing differentiation and specialization. In these descriptions, we find scholars that we can call 'separatists'. They not only analyse society based on ideas about differentiation into separate spheres, but also advocate for separation between the spheres – for example, between science and politics. Then, the focus of Chapter 3 is on models and ideas that show the practical implications of this separation. Here, we meet specific understandings of the interaction between science and politics, such as the linear knowledge model – which is based on the idea that knowledge precedes decisions – and the deficit model – which is based on the idea that laypeople lack the knowledge that experts have.

Part II, Overlap (Chapter 4), is an interlude between the other two parts of the book. It is devoted to approaches that seek a fruitful interaction between science and politics which can avoid the pitfalls of politicized expertise and scientized politics: the two threats of populism and technocracy. How should this be done? Different approaches to scientific experts and their relationships to decision-making are presented. We pay attention to all the work required to keep science and politics separate, but also make them communicate in a productive way. We describe boundary organizations and extended participation, both of which exemplify the overlaps between science and decision-making.

Part III, Co-production (chapters 5, 6, and 7), assumes that science and democracy are fundamentally interconnected. Co-production is a key concept in STS research; it means that science and democracy constitute each other and are created together. However, this co-production can lead to a number of different forms of expression. In this way, co-production includes separation as one possible expression of the interplay between science and politics. What becomes important in this approach are the *practices* of co-production. It is the acts of creation, formation, and construction – the co-production – that are studied. This means that the question of *how* co-production happens becomes of great importance. How is science and politics done? This *how* question then logically transfers us to the *who* question: Who does science and politics? The issue of participation understood as co-production is central in chapters 6 and 7. In Chapter 6, the focus is on participation in relation to technical decision-making in the public domain, while Chapter 7 addresses the notion of scientific citizenship.

Finally, in Chapter 8, we summarize the specific contributions of STS research to the discussion of science and democracy. We do this through the following four themes: representation and participation; separation and hybridization; situated practices and democratic theory; and STS and normativity.

PART I

Separation

2

Science and Politics as Separate Domains

Introduction

Science and democracy have emerged as important institutions in modern Western societies. But what do we mean when we say that a society is 'modern'? Among other things, modernity means that the world is understood in the light of scientific knowledge instead of traditional knowledge and religious beliefs. Scientific knowledge takes precedence over other forms of knowledge and becomes the yardstick from which other knowledge claims are judged. This situation becomes an important part of the characterization of modernity (Beck 1992: Chapter 7).

That science is superior to other forms of knowledge in modern society is only valid on a rather basic level. However, we are expected to accept that the earth was not created in seven days, that all living things consist of one or more cells, and that the dropped coffee cup falls to the ground due to gravity. The limit, or the boundary, of what issues can legitimately be answered from a scientific point of view can never be strictly formulated. Does climate change have human causes? Has extreme weather become more common due to a higher global average temperature? Is a global climate tax the best measure for reducing greenhouse gas emissions? Do we all need to change our individual lifestyles due to climate change? Where in the climate discussion does science end and politics start, and is climate change mitigation a scientific issue at all?

These questions illustrate that it is not obvious what questions science can answer and when science should have priority over other kinds of knowledge. The boundaries between science, values, and political assessments are not sharp in these kinds of issue complex (Hulme 2009). Many issues in relation to topics like climate change are both understood and managed by a mix of scientific knowledge and political assessment. Climate change would not be an issue at all if there was not an underlying value-based assessment that it is of interest to protect planet earth for future generations of human beings.

Modernity and its strong focus on scientific knowledge can be understood from the notion of *separation*. The anthropologist and STS scholar Bruno Latour (1993) characterizes modernity as a separation of nature (non-humans) from society (humans). This division is fundamental to modern societies and has led to a sharp division between science and politics. In the modern, enlightened society, both science and democracy are highly valued. They have, as discussed in Chapter 1, common roots in a critical public discourse which takes neither epistemic nor political authority for granted. However, they are often understood as founded in different logics, which motivates a strong separation. Science should not be politicized or democratized. Science works in an elitist way, which implies the necessity of separating those who know more from those who know less. Democracy, for its part, works in an egalitarian way, which implies that citizens are acknowledged as being equal and as having the same rights.

Latour questions the whole idea that modernity means a clear shift from 'an archaic and stable past' based on tradition and religious beliefs (1993: 10). Still, he understands separation as one of the basic principles of modernity, which is something he shares with social theorists who explain the emergence of modern society as a matter of increased differentiation and specialization. To this acknowledgement that separation has occurred, Latour makes an addition: there is no modernity and there never has been. This implies, according to Latour, that *separation* is a secondary phenomenon based on the essential foundation of *integration*. The world is one and holds together; what is divided and separated never gives the full story. In later chapters (see Chapter 5 in particular), we refer to these insights about a contingent whole as *co-production*, which is of special importance for understanding the mutual dependency between knowledge and social order (Jasanoff 2004).

This chapter focuses on the idea of separation and descriptions of how science and politics differ from each other. However, we also introduce scholars, such as Latour and Michel Callon, who are critical of this way of thinking while simultaneously acknowledging the existence and great importance of separation as a *social fact* with important consequences. In the third part of this book, titled Co-production, this duality is fully developed.

In the next section, we present social theorists who understand the development of society as increased differentiation, and we pay particular attention to those who talk about a clear separation between science and politics. In the section after that, we introduce a discussion of society as constituted by *autonomous social spheres*, which is an important understanding of modernity. Then we discuss the sharp distinction between those who *have knowledge* and those who *lack knowledge*, which is a consequence of the separation of and the authority given to scientific knowledge. This we do with the help of Plato's classical allegory of the cave. According to Latour, this allegory captures the crucial dividing line in Western society between those who know and those who do not know, including the belief that this difference is inevitable. In the final section, we return to the question of *representation*. With help from Callon, we explain how representation can be understood through the concept of *delegation* and further how the democracy we know today can be called *delegative democracy*. Delegative democracy is characterized by the fact that power has been delegated to a few, in terms of both political and epistemic power. This results in a democracy based on separation in relation to both power and knowledge.

All in all, this part of the book, focusing on *separation*, is an important and necessary background for the two later parts on *overlap* and *co-production*. Although we argue for a co-productionist approach, we also argue for the importance of acknowledging the separatist views and, not least, *separatist practices*. As becomes clear later in this book, we argue that separation could be better understood within the frame of co-production. However, a first step is to understand separation according to its own conditions.

Societal development as increased differentiation

A central theme for classical sociologists was the development of society as a process of increased complexity and differentiation. This is understood as a transition from a premodern to a modern society. The former is characterized by a low degree of division of labour and specialization, while the latter is based on social institutions being separate from each other and functioning according to independent logics. Sociologists have studied how institutions – such as family and kinship, knowledge and education, religion, markets, voluntary associations, and the state – have moved towards increasing independence linked to the emergence of specialized skills and professions. Among the founding fathers of sociology, Émile Durkheim (1858–1917) studied society's increasing division of labour at an overall level and Max Weber (1864–1920) analysed differentiation between several value spheres (Weber 2004). Today, we talk generally about the different logics of state, market, and civil society in terms of goals and ways of functioning. However, these spheres are all crucial for modern society, and therefore they must find specific and separate spaces for their activities. What constitutes modern society is manifold institutions with independent logics (Stichweh 2022). This is what differentiation is about.

For both Marx and Durkheim, and several others, the increased division of labour (specialization) is about resources (such as capital, education, competences, leadership). A complex society is a rich society where resources are accumulated and can be distributed. Separation understood as societal differentiation and group specialization is not a natural law, but part of societal development, even though Weber (2001) emphasizes the obvious risk of being captured in the 'iron cage' of taking increased differentiation for granted as an inevitable power in society. However, this is never the case, and since the founding of sociology, the basic task has been to critically study how specialization is performed and the consequences of this.

Among more recent scholars, Niklas Luhmann (1995) has developed systems theory, a sociology in which the differentiation of the social system means that social subsystems are functionally separate and integration between them does not exist. The unity of society is nothing more than the differentiation between the subsystems. According to Luhmann, this situation is explained by the evolution of the systems, which is about reducing complexity. These reductions can be done in different ways, and this leads to subsystems. In Luhmann's theory, the basic logics of the subsystems are strong, but they are the result of historical development and, therefore, can be changed, though this is not easily done.

Modern society is a differentiated society. Increased differentiation means increased political complexity, which in turn leads to problems for both knowledge and democracy. One reason for this is that various tasks in modern society are handled by relatively autonomous subsystems. How then can understanding across and coordination between subsystems be achieved? Subsystems that increasingly have the characteristics of 'expert systems', including highly specialized knowledge, pose a challenge for democratic societies. This is understood by Luhmann (1989) as a communication problem.

We find the idea of separation far back in history and also in philosophy. Since we are most interested in the relationship between science and democracy, we now address this in more detail and from the viewpoint of differentiation. An important distinction in classical philosophy is between facts and values, between 'is' and 'ought'. For example, Immanuel Kant (1724–1804) granted all citizens equal competence to decide on value issues, but not in relation to facts. This implicates that 'ought' can never be derived from 'is'. Trying to base values on empirical facts is called 'the naturalistic mistake'. Values are linked to regulation of actions, while this is not the case with descriptions of empirical conditions. Facts have no direction and do not specify goals. This more philosophical way of reasoning means cutting the link to society: separation becomes a question of universalized ontology and not societal development.

From this reasoning, regardless of philosophical and sociological positions, science and politics become two distinct spheres, governed by different dimensions of validity: *truth* and the *power to make binding decisions*. Describing a factual relationship is something completely different from describing what should be done and how. However, both descriptions may be valid, though on different grounds. From this, two different professional roles emerge with distinct tasks: the scientist and the politician (Weber 2004).

The distinction between a value sphere and a factual sphere is important for understanding Weber's critique of the increased rationalization in modern society. According to Weber, a dominant form of rationality is a rule-driven bureaucracy, which is strongly focused on developing rules, procedures, and technical solutions. Therefore, we need to think more deeply about what scientific development and technological progress entail. Weber focuses on the risk that the modern citizen will neither have a better life nor become a more developed human being. Such a feared tendency implies 'specialists without spirit, sensualists without heart; [and] this nullity imagines that it has attained a level of civilization never before achieved' (Weber 2001: 124). The factual sphere risks colonizing the value sphere, and for the worse.

For Weber, politics and political action belong to a value-based sphere. Goals are never given but must be formulated, chosen, and evaluated. What values we should live and strive for cannot be determined by rational considerations. Political action cannot be rationally substantiated. Rational action is simply not applicable when it comes to deciding about values, goals, and needs. What happens in a rationalization of politics, and other spheres of society, is that choices themselves become rationalized, which means that technical solutions become goals rather than means. Weber calls this world that hides the value sphere 'rationalized', and (later) Jürgen Habermas (1971) calls it 'scienticized' and 'technocratic'.

Weber's critique of increasing specialization and rationalization in Western societies was an important starting point for the Frankfurt School's critique of modern technocratic society. Technological progress, together with mass production under capitalism, is an important basis for the dominance of instrumental reason. The abundant presence of technological artefacts means that we use them without asking ourselves why. In the field of politics, we see that instruments for prediction and control are considered necessary and are spreading into more areas (Jasanoff 2003). Habermas talks about a technical interest gaining increasing influence over social institutions. Understanding and judgment are set aside, and from a scienticized politics follows a depoliticized public.

In the book *Dialectic of Enlightenment* from 1944, Max Horkheimer and Theodor W. Adorno draw these thoughts to a dystopian end. In this book, written in the US while the authors were in exile from the German Nazi regime, they try to grasp the future of the Western world. For Weber and the Frankfurt School, the interplay between values and facts in the modern world has led to a tragic situation of increased rationalization.

The idea of a strong separation between science and politics, and a more general functional differentiation in society, has dominated both social philosophy and sociology. Talcott Parsons' structural functionalism, which had a major influence on American and European sociology for many decades during and after World War II, is an important example of this dominance (see Trevino and Staubmann 2022). Along with Parsons, Robert Merton was a leading representative of this tradition; Merton was the first sociologist to study science as a social institution, and he also founded the sociology of science as an academic specialty. In accordance with the functional differentiation of society, Merton distinguishes science as an autonomous social institution.

According to Merton, science and democracy are two different and independent institutions. Merton's idea is that a social institution is constituted by a set of specific norms that distinguish it from other institutions. Science is an independent institution as long as its basic norms control the actions of those who work within it in a way that distinguishes it from other institutions. In other words, an institution must be able to present specific norms that are adhered to. The four norms that regulate the social institution of science are summarized by Merton (1973a) in 'the scientific ethos': universalism, communism (often presented by today's scholars as 'communality' due to the political connotations of 'communism'), disinterestedness, and organized scepticism.

However, the autonomy of science has, according to Merton (1945: 415), a clear downside, since frustration arises when research is to be used in decision-making and when researchers collaborate with representatives from other institutions. This is the problem of communication across institutions we referred to earlier.

Merton's conceptualization of science as an independent social institution has had huge influence on the social sciences and, more generally, in public discussions of science. The norms of universality and disinterestedness are widespread. Few deny that science should be independent in relation to class, gender, and ethnicity, and not governed by private (individual as well as organizational) interests. Violation of these norms means that science becomes less scientific – that is, the social institution of science becomes less independent.

The idea of science as an autonomous institution is strong. Even among feminist scholars who criticize science for systematic misrepresentation of women, we find those who believe that these are inequalities that can be remedied by a better science, performed in a true Mertonian spirit (Longino 1990: 11). Science is believed to be able to deal with its own problems in an autonomous way. Political interventions in science are therefore often condemned in modern societies as illegitimate – not least by scientists, who are keen to protect their own independence and that of the institutions they represent, and Merton's norms are often used strategically to further this aim (Gieryn 1983).

But what is the meaning of these philosophical and sociological postulates about separation and autonomy, which often form the basis for motivating and justifying science and for maintaining strong boundaries between science and politics? For now, we should do nothing more than conclude the importance of noting the status of the areas which are separated and the boundaries which perform and protect this separation. How are they motivated and explained? Are they given a fixed ontological status, or are they understood as being the result of a specific historical and societal situation? Our standpoint is sociological, which means we understand separation as societal differentiation and group specialization, and as historically changing. This implies that questions about *what* to separate and *how* are open for discussion, from a practical as well as from an analytical perspective.

That separation is a question for discussion is an idea easily found in recent social science literature. Here, we find many examples of those who criticize the autonomy of science in terms of both ideals and reality. For instance, Helga Nowotny and colleagues (2001) have developed an ideal-type distinction between research performed separately from politics and research carried out in a collaborative way together with political actors; these are called Mode 1 and Mode 2, respectively. The aim here is to highlight the interplay between science and society as being not only about separation but also about integration. According to Nowotny et al, and thus in line with Latour's argument which we met in the introduction of this chapter, integration is primary and separation secondary.

The two positions of separation and integration are often presented as existing simultaneously, and this is especially visible when policy-relevant research is aimed for (Sundqvist et al 2018). This research is about solving societal problems without being politicized. The involvement of science in political decision-making can be presented positively, as socially relevant and useful, or negatively, as politically influenced and controlled. We find conflicting views on how to assess this research, but also ambitions to both eat one's cake and have it, to defend both autonomous institutions and close cooperation, separation as well as integration. However, our simple argument is that separation is – and should be – open for discussion and critical analysis.

Autonomous institutions

The American philosopher Michael Walzer (1983) has formulated a theory of democracy that tries to solve the previously mentioned 'communication problem' across independent social institutions. This theory can also be used to criticize abuses of democracy, such as technocracy, as being about illegitimate expert rule. Walzer is in favour of differentiation and assumes that modern democratic societies are divided into several basically independent spheres, which are functionally differentiated. Each of these institutions has an important function for the reproduction of society. Walzer goes as far as to call them autonomous. What is perceived as a fair and valid decision-making order within one sphere cannot, therefore, be transferred to another. This order simply does not work and is not recognized as valid in other spheres. Walzer's theory is an interesting example of how the boundaries between functionally differentiated spheres can be understood and, not least, how coordination between them can be performed.

Without claiming completeness, Walzer (1983: 318) enumerates six spheres: political power – the state; law – judgments; economy – money and commodities; knowledge – education and qualifications; family – kinship and love; and health and care – security and welfare. When the principles of decision-making, distribution, and justice from one sphere penetrate and become governing principles also for other spheres, dominance arises, and according to Walzer (1983: 10f), dominance is always illegitimate. Dominance deprives the autonomy of the spheres. For democratic societies, however, there is one important exception, and that is the legitimate dominance of the political sphere over the other spheres. In a democracy, the political sphere is superior. This does not mean that political principles – such as majority decisions – are also applicable in other spheres, but only that political power has the right to define the boundaries of the other spheres. Walzer (1983: 15) expresses this as politics being dominant at the borders of the spheres but not within them. Politics sets (and should set) the framework for all activities in society.

Political power has the right to restrict how individuals decide over their own lives – for example, by designating where people are allowed to drive cars in the city and where they can park. Some restrictions set on car use today are motivated by climate change and aim to persuade people to move from driving cars to using public transportation, riding bikes, and walking. However, the restrictions are not (at least not yet) about the general question of car use. As long as car use is allowed in principle, regulation will concern more specific things in relation to this, such as the required training, the need for sobriety, the use of seat belts, the requirement to adhere to speed limits, and the need for car safety. This simple example is about politics setting boundaries for individual actions, without domination.

According to Walzer's view of democracy, the political sphere has the right to decide the boundary between science and politics. How this boundary is defined cannot be anchored and legitimized in any way other than through the common values possessed by the people within a certain culture. The spheres have no intrinsic values, and the boundaries between them cannot be formulated universally by philosophers or scientists. It is not a factual issue. This means that there is no other basis for legitimate demarcation – and for criticism of existing boundaries – than that which can be attributed to commonly held values. If there is public acceptance that experts should decide on a certain issue, this is not technocracy. The same decision by the same experts can in other circumstances be characterized as technocracy if it is a matter of illegitimate expansion of one sphere at the expense of other spheres – that is, dominance. The question of what is legitimate and illegitimate expert rule, or more generally what are political and scientific issues, is thus dependent on how the political sphere, based on democratic decisions, has discussed and carved out the boundaries between the spheres at the time. In Walzer's theory, common values form the cornerstone and a base for society as a whole. Both truth and justice, and not least the limits of their extent, must be blessed by common values. One important conclusion from this theory is that science and its practice rests on public trust.

Societal worlds are also of importance for STS theories, especially those inspired by Niklas Luhmann's systems theory. In this vein, Reiner Grundmann and Simone Rödder propose a 'multiple-worlds model', which assumes that 'the worlds of science and policy follow rather different institutional logics: politicians adhere to scientific advice for political reasons, or they do not adhere to it for political reasons. And scientists strive to find evidence for their theories and hypotheses, even if they are politically inconvenient' (2019: 3886). Communication across the worlds of science and policy is possible but difficult, and this cannot be solved by more and better knowledge. What is needed is an awareness of the different logics of the separate worlds, which interpret knowledge in different ways for different reasons. The multiple-worlds model 'calls for an acknowledgment of these communicative risks as inherent to science-policy debates' (Grundmann and Rödder 2019: 3886).

Along similar lines, Peter Weingart (1999: 155) argues that in today's Western societies, scientific knowledge is widely distributed and has strong legitimacy due to its epistemic authority. Especially concerning risk issues, scientific knowledge is of crucial importance in defining problems and influencing political agenda-setting. This importance of scientific knowledge also implies that it is increasingly playing the role of legitimating politics and, thus, a scientization of politics. These opposing but connected tendencies in science advice are growing increasingly stronger in today's societies as more knowledge of different kinds is funded, produced, and spread. But instead of making this situation the basis for theoretical considerations, Weingart argues, using ideas from Luhmann's system theory, that scholars who want to improve science–society interactions should not be trapped by this situation of increasing mutual influence between science and politics. To not acknowledge the differences between science and politics could be viewed as adding to the problem.

Weingart's position is that the interaction between science and politics – which proliferates in modern societies due to the increased use of expert knowledge – is a key object of study for social scientists. However, these activities, according to Weingart, are secondary to the basic logics of the two spheres of science and politics (Maasen and Weingart 2005). In short, science and politics are changing but should be acknowledged as being more stable in their differences than the many collaborative activities in which they are combined. This argument is different to what has been proposed, as discussed earlier in this chapter, by Latour, Nowotny, and others who argue that integration and collaboration are primary and separation is secondary.

In the field of climate change governance, we also often find what we can call *separatist models*. Social scientists studying science–policy interactions in relation to climate change often start by identifying two separate worlds, but also attribute problems to such separation (Sundqvist et al 2018). A problematic gap between science and policy is described. Greenhouse gas emissions increase at the same time as knowledge about this problematic situation becomes more alarming and more widespread. The knowledge–action gap is widening and is often talked about as a communication problem. Words like 'barrier', 'obstacle', 'hindrance', 'constraint', 'hurdle', and 'tension' are often used (see, for example, Eisenack 2014), and the situation is understood as being about *problems* or frustrations, as Merton talks about, that arise when the two worlds meet. The gap is the reality, and 'bridging', 'linking', 'shared understanding', 'dialogue', and 'interaction' are the proposed *solutions* to situations when the gap is seen as problematic (Mastrandrea et al 2010; Dilling and Lemos 2011). However, as in the multiple-worlds model, the basic understanding is based on science and policy as two separate worlds. A separatist view stands strong in the discussion of climate change issues and how to manage them, but scholars diverge when it comes to how to understand and assess this situation. So far, in this section, we have shown that this is the case more generally among social scientists attempting to understand and assess science–policy interactions.

Plato's cave and Latour's two chambers

In ancient Greece (700s BC to 300s BC), the first thoughts about and also the first practices of what we today call a 'democratic system of government' were formed. In the Athenian city-state, all free men over the age of 20 whose father was an Athenian citizen had the right to participate in the People's Assembly meetings. In addition, participation in political life was strongly encouraged. It was not only a right but also a duty to participate in meetings where things of importance to the city were dealt with (Fustel De Coulanges 1980: 213). It wasn't only political activities that were of great importance, but also philosophy. Reason was highly valued, though in a way that we today would find almost naive, since reason was assumed to govern action. In addition, Plato, the most famous of the ancient Greek philosophers, believed that philosophers should also rule. In this sense, ironically, Plato can be considered an anti-democrat and a technocrat. Plato formulated a sharp distinction between knowledge and politics, and not least a strong boundary between those who know and those who do not know. In this respect, he argued for separation.

Plato formulated his well-known cave allegory to illustrate the sharp division between nature and society and between different kinds of knowledge. The allegory, found in *The Republic* (Plato 2007), tells of a group of people sitting chained in a cave, secured around their legs and necks so that they can only see straight ahead. Behind them a fire is burning, and between this and the prisoners there is a road with a low wall running next to it. On the road, people are carrying tools that can be seen above the wall. With the help of the light of the fire, the movements of the people and the tools are reflected as shadow images on the wall in the cave, to which the eyes of the chained people are directed.

The allegory goes on to tell of the hypothetical scenario in which one of the prisoners escapes by passing the wall and the fire, reaching the opening of the cave. This cave dweller then, on his way out, is able to understand that the figures on the wall were sham pictures that he can only now interpret correctly. However, once out of the cave, the sun dazzles the prisoner so the chances of seeing anything at all become small. What Plato depicts are the difficulties and pains associated with leaving the cave and seeing the light. After a while, however, the prisoner becomes accustomed to life outside the cave and convinced that in the sunlight the world can be seen in a truer way than is possible for those in the cave, who see shadow images which they perceive as reality.

According to Plato, the prisoner, who returns to the cave after seeing the light of the sun, would find it difficult to accept the darkness and the shadows, and at the same time would also realize the difficulties of convincing his former fellow prisoners of the light on the outside and that what they saw in the cave were shadows. It is, thus, no easy task to go from darkness to light and from light to darkness.

This allegory, which is one of the most well-known in Western philosophy, is a summary of the fundamental relationship between nature and society and the hierarchy of different kinds of knowledge. The prisoners in the story are ordinary people trapped in minds that have insufficient ideas and knowledge of reality, and therefore they are content to believe illusions. The real, or higher, knowledge (the world of ideas) is difficult to access, as to reach this, one must leave the common world. According to Plato, this is what philosophers do, but their knowledge is not only difficult to acquire but also difficult to teach and learn, and not least it meets resistance and criticism when communicated to ordinary people.

Latour (2004) uses Plato's cave allegory to illustrate the strong division between science and politics in the Western world. Latour describes modern society as based on a two-chamber parliament: one represents facts/nature and the other people/society. The two-chamber system is based on a sharp boundary, which means that the representatives of one chamber (science representing nature) cannot easily go between or over to the other chamber (politics representing the people).

Nature cannot speak for itself, but needs someone who can represent it in order for knowledge to be produced. In modern societies, these representatives are called 'scientists'. Nature belongs to all the things that are not human and do not belong to society (planets, rocks, water, plants, animals, cells, molecules, atoms, and so on). However, the sharp division between nature and society leads to paradoxes or mysteries. If the difference is so sharp, how can people (from society) move into nature and understand it? And how can they then come back and represent nature to people in society, and make themselves understood (Latour 2004: 13–14)? The two transitions, the entrance and the exit (from the sun to the cave, or from nature to society, and back again), which are about bringing truth to society, have remained a mystery since the time of Plato. We need to turn myth into reality and understand actual processes of the interplay between nature and society and how knowledge is produced and communicated.

Latour describes this mystery as a situation where scientific experts enjoy the most amazing political power. These representatives of science automatically become representatives of nature and things. This small group of experts 'can make the mute world speak, tell the truth without being challenged, put an end to the interminable arguments through an incontestable form of authority that would stem from things themselves' (Latour 2004: 14, emphasis removed). The strongest force that we find today in Plato's allegory of the cave, according to Latour, is in the idea of knowledge – true knowledge – as something that stands above and outside the social world. This is where the defence of a two-chamber parliament, and the asymmetric distribution of power between them, stands out most strongly.

The irony of the fact that both science and democracy were born in ancient Athens is that ever since Plato's philosopher returned to the cave to educate the prisoners, scientific representations have been used to silence and look down on democratic discussions (Brown 2009: 169). In Latour's words, 'representation ... rendered democracy powerless as soon as it was invented' (2004: 71). Scientific experts become a protection against the constantly imminent threat that the ignorant masses take over if knowledge is not given priority over democracy.

Latour's reinterpretation of Plato's cave allegory is interesting for two reasons. First, Latour makes it clear that the relationship between those who know and those who do not know is conditional and about codependency. This means that Latour is adding a sociological explanation to Plato's universal presentation. Second, and consequently, expert authority becomes central but is always in need of explanation. Separation could exist, and quite often this is the case, but we should never take it for granted, and we need to explain it instead of accepting it as a mystery. Yet separation gives no clues on how to proceed in support of this aim.

Delegative democracy

Michel Callon, together with colleagues, refers to the strong division between science and politics that Latour describes as *delegative democracy* (Callon et al 2009: 35, 119–120). This democracy is based not on one demarcation but two, both strongly defended in Western democratic societies: the boundary between voters and their elected representatives; and the boundary between laypeople and experts. In the overall division between science and politics, we find a double delegation: political issues are delegated to elected representatives; and technical issues are delegated to specialists. The similarity between these two divisions is the idea of representation, in the sense that issues are delegated to a deputy elite who can act as represents. The two elites of politicians and specialists form the basis of delegative democracy. The two demarcations lead to one and the same consequence, namely the establishment of an undifferentiated passive and ignorant public, who are not considered competent to participate in decision-making on either political or technical issues (Callon 2009: xx). According to this idea, it is not only the scientific elite that has succeeded in leaving Plato's cave to seek the light but also a political elite, while the great mass lives in the darkness of ignorance in relation to both knowledge and power.

Delegation to specialists means that knowledge is produced in secluded locations and by a few. These specialists have succeeded in scientifically problematizing a selected part of reality, simplifying it and making it technically manipulable. Groups of specialists form around specific ways of problematizing. The uncertainties that arise in this work are regarded as issues that can only be handled by technical experts, who have the right to form cognitive agreement in isolation from the rest of society (Callon et al 2009: Chapter 2).

A prerequisite for the specialists to be able to cultivate their development of knowledge in isolation is that they can convince the political elite, or an industry elite if this is responsible for giving mandate and financial support, that the technical problematization can lead to something important and desirable, such as healthcare technologies that save lives, or to the production of high-yielding goods, such as nuclear power plants generating electricity. In these deliveries, we find the epistemic power and the high status which scientific knowledge (and the technology that follows in its footsteps) has acquired in modern society. It is with hopes and promises of improvements based on technical solutions that researchers and engineers gain authority. It is they who, with the light in their hands, can reform and improve our cave life; with the help of new energy sources and medicine, we can all live better and brighter lives. A technical problematization of social problems means a promise that what was previously done manually, or could not be done at all, can now be done with the help of mechanical, electronic, chemical, or nuclear aids. In this way, science provides improved prostheses for our daily lives (Barthe et al 2022).

Delegation is about what issues are (or can be) delegated to technical experts and political elites, who are trusted to deal with how nature and society are understood, managed, and taken care of. In cases where there is established and legitimized epistemic and/or political authority, delegation is considered uncontroversial and also desirable because competent groups can deal with and solve our common epistemic and political problems. But when authority is uncertain and disputed, delegation becomes controversial and governing elites' way of doing things becomes contested. We should therefore talk about degree of delegation and not least about the possibilities of recalling a given delegation.

The degree of delegation varies between different countries and between different types of issue. France, where Callon and his colleagues have carried out studies, is a more elitist country than, for instance, any of the Nordic countries, which have a more decentralized democracy in the form of municipal self-government and a long tradition of civic participation at the local level. On issues that are controversial – in a scientific and/or political sense – we can expect delegation to be called into question. In such circumstances, when uncertainties and complexities arise in broader societal discussions and within groups outside the two elites, it becomes more difficult to justify that issues are being handled by small, secluded elite groups.

To conclude, Callon and his colleagues, through a focus on the two delegations and the boundaries developed to motivate delegation, present a useful map for further empirical studies on how expertise works in relation to politics in a democratic society. There are no ontological essences to help us understand or navigate this map – that is, what to study and what to focus on in relation to issues, actors, responsibilities, and interrelationships. This will, rather, depend on each situation, and where, and about what, agreements or controversies arise at a given point in time.

An important example of delegative democracy, and one where delegation has been questioned, relates to nuclear power (see Callon et al 2009). In Sweden, as a prominent example, there was a national referendum in 1980 on the future of nuclear power (Sundqvist 2002, 2012). This could be understood as technical and political delegation taken back to the people. When nuclear power was developed in Sweden, shortly after the end of World War II, the political elite, in close contact with nuclear scientists, decided on huge investments in the peaceful development of this technology. The public was not involved and, in principle, had no insights into these plans. The government established an advisory body of scientific experts, which soon was given the financial and organizational resources to develop the peaceful use of nuclear power through a state-owned company. With promises that this was the energy source of the future, which would produce clean and cheap energy, the (ignorant) public had no reason to question the delegation to the two elites in close symbiosis.

During the early 1970s, when the Swedish nuclear power programme was in the middle of its development, something happened. When six reactors were in operation and another four were ready to be fuelled, nuclear power started to be questioned intensely by the environmental movement and by groups within political parties in the national parliament, but also by individual specialists and concerned citizens. Critics argued that nuclear power involves several risks that the elites had ignored or kept secret. The technology also proved to be significantly more expensive than planned and produced hazardous waste that would remain so over long periods, and for which there was no plan for disposal.

During the late 1970s, nuclear power was the dominant issue in Swedish politics and public debate. It was discussed by all, drawing on both scientific and political arguments. The elites were uncomfortable in their roles, as they now strongly felt that their delegation was threatened. The political elite recognized no way forward other than to call a national referendum on the future of nuclear power. Many of the scientific elite were upset and believed that holding a referendum on such a technical and complicated issue was unreasonable. The result of the referendum also turned into a controversial compromise due to unclear alternatives in voting on, first, continued expansion and, second, a phase-out process. What this example mainly shows is that the delegation of nuclear power was withdrawn, and since then nuclear power has been a controversial issue with clear problems of legitimacy regarding delegation to the technical–political elite. The ignorant public became less ignorant and very much engaged.

The Swedish nuclear power debate was part of an international pattern. In particular, critical arguments against nuclear power were imported from the US, where a debate had arisen earlier. This was exacerbated by an accident in a reactor at Three Mile Island in 1979. Similar debates could be found in most Western democracies with nuclear power programmes at this time. The debates inspired each other, but national and local variations should not be underestimated. The differences have to do with trust in experts and politicians as well as the strength of the environmental movement and civil society, and their organizational resources (Rudig and Flam 1994). In other words, there are national and local differences in how global or European trends are staged by different actors and how they are translated into national and local practices.

The nuclear debate is interesting because it was the first major popular debate on an issue that was considered by most people at an earlier stage as being legitimately delegated to technical experts. The issue formed a pattern for later debates on chemicals, genetic engineering, stem cell research, artificial insemination, food security, nanotechnology, climate change, and artificial intelligence. These issues are all scientific but are also in the wake of nuclear power seen as ethical and political, since they are about what sort of society we want to live in, how risks and benefits should be weighed against each other, and not least what role the coming generations should be attributed (Felt 2015: 105f). Consequently, they are problematic to deal with, both scientifically and politically, which means that representations are questioned.

Callon (2009) believes that nuclear power has been crucial in creating and maintaining a separation between the political and the technical – that is, the boundary between voters and their elected representatives and the boundary between laypeople and experts. The previously mentioned issues with nuclear power at the forefront have arisen on the basis that a scientific–technical elite in isolation and often with mandate and support from the political elite has developed new technical possibilities, which they want to further develop and to which great promises are attached.

The issue of climate change is both similar and different to nuclear power. Delegation to experts is highly visible. The Intergovernmental Panel on Climate Change (IPCC) has been a distinctive and authoritative United Nations (UN) expert body since 1988, and it regularly, in cycles of six or seven years, delivers comprehensive scientific assessments of the climate situation. These reports present a global scientific consensus on what research says about climate change, its consequences, and possible solutions. From the perspective of delegative democracy, the IPCC could be interpreted as an isolated scientific elite. However, so far, climate experts have not been used by the political elite in a way that is comparable to the example of the nuclear experts. Moreover, the IPCC itself does not seem to be interested in producing usable knowledge, since they work from the self-imposed mandate of being policy relevant but never policy prescriptive. Maintaining a sharp dividing line between its own work and the domain of policy and politics is an important aim for the IPCC (Hermansen et al 2021; De Pryck and Hulme 2022). Today, many scholars, and not least the climate movement, are focusing on the knowledge–action gap and how delegation to experts is part of the problem. Does this mean that delegation to experts to present the truth about climate change has become too strong and should be recalled, or just that it should be reconfigured? In any case, it is obvious that the science is separate from the politics and that the knowledge presented is not enough to require particular actions.

In the cases of climate change and nuclear power, we can ask questions concerning how sustainable existing delegations are – for example, regarding how responsible and far-sighted the experts are regarding issues of societal

Figure 2.1: Delegative democracy as separate representation

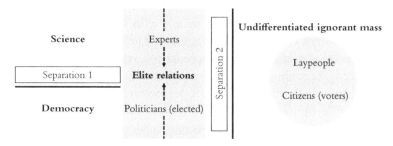

implications, long-term uncertainties, and global justice. In short: what do the blind spots of experts and elites look like? What we have shown in this chapter is that delegation exists and is often in stable configurations of double delegation – that is, based on clear separation of science and politics – but also that delegation can be questioned, sometimes so forcefully that it is taken back by the people.

We can now summarize delegative democracy, making visible two basic separations (Figure 2.1). The first is a separation between the worlds of science and democracy, and the second follows from this by distinguishing the two elite groups of scientific experts and politicians on the one side and the undifferentiated group of ignorant people (laypeople and citizens) on the other.

A consequence of these separations is a power asymmetry between the cooperating elites (the politicians and experts) and the ignorant people. Important to note, however, is the existence of empirical variation when it comes to how the two elite groups interact, and also how the two groups are composed and how they interact with groups of the public. In the example of nuclear power, we find applied research (scientific elite) in close cooperation with the state (political elite) and industry (technical experts) in addition to marginalized publics that turned into engaged citizens who recalled the delegation. The example of climate change presents a rather different picture of the relation between the two elites of experts and decision makers, in which the scientific experts (the IPCC) are marginalized in relation to the political elite. Such observations are a good start for detailed empirical studies as well as for comparisons on how different issues are managed.

Conclusions

In this chapter, we have met ideas about knowledge and democracy which are of fundamental importance for how modern society understands itself. Plato's cave allegory is a classic example. In this, the enlightened expert (the philosopher) should guide the ignorant. Callon's delegative democracy, unlike what Plato's allegory tells us, is also a recognition of the power of politics and democracy. Instead of *one* elite, we get *two*. But in contrast to Walzer, who believes that ultimately it is the people who set the boundaries for all activities in society, Callon's model suggests that *in practice* leading politicians and experts often agree about the boundaries, while the public becomes marginalized. According to Callon, the development of nuclear power, which began with the development of the atomic bomb, is a paradigmatic example that seems to still guide how advanced technological development takes place in modern democratic societies. Such technologies are often costly national prestige projects which require government intervention in order to be funded and realized. In this way, strong connections between universities, governments, and businesses are established, but without much transparency and opportunity for publics to engage and to have influence (as in the undifferentiated group of ignorant people who are marginalized in relation to both epistemic and political power). Plato's philosophers and Callon's description of the agreement between the two elites can both be seen as forms of elite rule or technocracy, which the Frankfurt School vividly described and strongly criticized.

Merton's description of differentiated and autonomous social institutions gives a non-technocratic contrast, since the autonomy of science does not imply that science dominates other social institutions. We also find this view among the many social scientists who study science–policy interactions in the environmental field and in relation to issues of regulation more generally. However, the autonomy and independence of science as a social institution may lapse if it does not maintain the specific standards and norms which characterize scientific work. For instance, strong industry connections mean that the norm of disinterestedness – which, according to Merton, is a cornerstone for science as a social institution – is threatened. Science becomes goal oriented and not independent. Like Merton, Walzer considers independent spheres as a positive way to maintain democracy and an enlightened modernity. Walzer also develops a democratic model for the relationships between societal spheres, based on non-dominance. Intrusion into the internal affairs of autonomous institutions is illegitimate domination. The idea of separation cannot be expressed more clearly.

Separation is often the case in a modern society, and no one is clearer about this than Latour. Based on his provocative slogan 'we have never been modern', empirical studies of this situation – including a focus on how separation, delegation, and autonomous societal spheres are upheld in practice – are proposed as an important analytical focus. This, in our view, is also the relevant STS position on how to deal with separation when studying science and democracy: to focus on the interplay and the different logics between domains separated in potentially different ways. When performing such studies, an important point, from our STS position, is that separation is *not* about ontology, but about specialization. However, the need and relevance of specialization is often sought from a practical perspective – division of labour is seen as something good by many actors – but can always be discussed, while an ontological understanding of separation closes down such discussions. An ontological understanding of separation thus closes down democratic debates. A perspective on separation as a *social fact* accepts that democratic debates might need to be closed down, but also that they might need to be reopened. How to assess when an opening up is needed and who can make such assessments have been reoccurring discussions within STS (Stirling 2008; Soneryd and Sundqvist 2022) and something we discuss in later parts of this book (chapters 6 and 8).

3

The Relationship between
Science and Politics

Introduction

In this chapter, we continue the presentation of ideas about science and politics as separate. While the previous chapter focused on separation as part of wider societal changes and presented thinkers who demonstrate a clear separation, some of which also support the idea of strictly separate domains, this chapter focuses on approaches that discuss how science and politics can and should relate to each other, while simultaneously acknowledging that separation exists. Many of the scholars discussed in the chapter study what happens when science and politics meet.

We first present the view that the increased importance of scientific knowledge for political decision-making has led to stronger demands for *scientific consensus*. For scientific experts to effectively impact on and influence political decisions, their knowledge base must be generally accepted among other scientists. We then describe the *linear model*, which is based on the idea that knowledge precedes action. The *deficit model* follows from the linear model and implies that the public is characterized by knowledge deficits, which can be remedied with education and reliable information. We then present Jürgen Habermas' seminal *pragmatic model* as a way to manage the gap between science and politics, and also the technocratic tendencies in modern society based on the increasing dominance of expert knowledge.

In the two final sections of this chapter, we introduce classical STS research on what happens when scientific experts become involved in political decision-making processes. Dorothy Nelkin's studies serve as an important example. Nelkin's conclusion is that politics frames the work of experts and how expert knowledge is used, and thereby expertise is reduced to a tool in managing political conflicts. Thereafter, we present Harry Collins and Robert Evans' characterization of the first, second, and third wave of STS research. In their approach, presented as the third wave, a clear separation between science and politics, and between experts and non-experts, is defended from the explicit aim of avoiding both technocracy (relying too much on knowledge) and populism (relying too little on knowledge).

Scientific consensus and autonomous experts

Peter Haas, a political scientist well known for the notion of *epistemic communities*, has for a long time studied international environmental governance and under what circumstances expert knowledge can gain political significance. He is particularly concerned with the question of how good decisions can be reached with the assistance of scientific knowledge, and scientific consensus is assessed as key to understanding why science will sometimes influence decisions and sometimes not.

When epistemic communities, as mediators between scientific knowledge and political decision-making, can refer to a larger scientific community that stands united on a certain issue, then expert knowledge will have good possibilities to influence politics (Haas 2007; see also Lidskog and Sundqvist 2015). In order to accomplish this, however, the experts must initially be separate from politics. It is only after an initial separation that science can have political influence at a later stage. Scientific practice must be left undisturbed, without political interference, in order to create a knowledge base that the scientific community can establish a consensus around (Haas and Stevens 2011).

According to Haas, it is important that expert knowledge is not linked to politics before scientific consensus has been reached, since science then would lose its autonomy. For example, a government should not select its scientific committees, but leave the selection process to scientists themselves, and make sure that the selection is based solely on scientific merits. The more autonomous science is, the more political influence it may have (Haas and Stevens 2011: 131). An example of this, presented by Haas and Casey Stevens, is the Convention on Long-Range Transboundary Air Pollution in Europe. The working groups and networks of expertise that contributed knowledge to the framework of the convention are all focused on identifying research questions and conducting independent research, and they have clear autonomy in relation to political decision-making. If scientific experts are not allowed to identify and formulate research questions independently, and as separate from political processes, scientific consensus will not occur and the results will, instead, be politically based compromises. The ideal relation between science and politics can, according to this approach, be seen as a process in three sequential steps: (i) science is kept separate from politics; (ii) scientific consensus is established; and (iii) consensual science is communicated to political decision makers.

The problem that Haas identifies in many efforts to involve scientific experts as advisors in political decision processes is that the involvement happens too early. Scientific expertise is included before scientific consensus is established. This eventually means that scientific controversies will come into focus, since there is no consensus, and the possibility for science to become a legitimate and trustworthy advisor is severely diminished. No learning will take place and political interests and divergencies will proliferate through this type of immature cooperation between experts and politicians. Haas and Stevens (2011) mention climate change governance, biodiversity, and fishery as examples in which expert groups have not received clear autonomy and thereby the possibilities for scientific influence have been undermined.

The UN's Intergovernmental Panel on Climate Change (IPCC) is, according to Haas and Stevens, an example of lack of scientific autonomy. When new scientific assessments are planned and organized, leading to IPCC reports, politics is involved in the work from the beginning, if not dominating the work. The IPCC is led by governments who decide about and organize new assessments. National governments nominate authors for the assessments and also approve the reports. According to Haas and Stevens, this means that already from the beginning, involved experts need to adapt to political will and, thereby, expert knowledge loses its potential to surprise decision makers. Politicians get what they want and expect from experts (see also De Pryck and Hulme 2022).

Haas' approach to epistemic communities comes close to what has been called the model of *speaking truth to power* (Sundqvist 2019). This approach concerns how to mobilize the best available knowledge for political decision-making. Experts believe they have something important to tell decision makers that the latter are not yet aware of. To surprise decision makers and to give them something they have never asked for, but is in great need of – once they understand – is what this model is about. In practice, however, we know that unwanted advice is never easy to give and often it is meaningless when power does not listen.

Haas' approach is influential in research on science–policy interactions concerning international governance of environmental issues. It is based on a clear separation between science and decision-making, and most of all it tries to answer the question of under what conditions expert knowledge could influence political decision-making.

The linear model

Speaking truth to power implies that knowledge is important and that decision makers do not have access to knowledge and therefore need to learn from others – the experts. This thinking is similar to the well-known *linear model of knowledge*, according to which scientific knowledge can influence, or even force, actions and political decision-making (Jasanoff 2017). The linearity means that scientific knowledge comes before, and as a prerequisite for, decisions: no knowledge, no action. Silke Beck (2011: 298) summarizes the model in three propositions: (i) more and better science will lead to better decisions; (ii) more and better science can solve political disagreement; and (iii) more and better science can make politics more rational and evidence based. In short, this means that politics should be science based. Before a political decision is taken, it should be asked what science says. This model assumes that science is neutral and independent from politics: science precedes politics and can therefore not be influenced by it. Haas' ideas about epistemic communities and the importance of scientific autonomy, described in the previous section, provide a clear example of the linear model of knowledge.

The linearity from knowledge to action has its roots in a linear *innovation model*. One strong assumption behind ideas about innovation is that wider developments in society are knowledge based (Bijker 1995; Durant 2016: 19). The process of innovation starts with basic science, which then follows a chain of logical steps, such as applied science, pilot-based production, mass production, wide societal use, and economic growth. According to this model, basic science is the engine for societal development and thereby also motivates political investments in science to promote a positive societal development. We can easily recognize this idea in political debates and in the argument, common in developed nations, that investments in research is a building block of both economic growth and welfare development.

There are several important implications of the linear knowledge model. In its strongest version, the political domain will be transformed into an arena for implementing research results; politics will equal following the advice of scientific experts. This means technocracy realized. A weaker version means that scientific knowledge becomes a central part of the political discussion and will influence and govern political debates (Beck 2011: 299). In both cases, the linear model of knowledge focuses on the importance of knowledge for political action, which means a reduction of the autonomy of the political domain. A pluralistic and conflictual political debate is seen as something unwanted and which can be gotten rid of through unambiguous and certain knowledge. This is what Habermas (1971) terms a 'scientization of politics'.

Climate change is an interesting example in this context. Several STS researchers have described the IPCC and its assessments of the state of knowledge in climate research as organized according to a linear knowledge model (Jasanoff and Wynne 1998; Beck 2011). There is a wide agreement that climate change is an issue that is dependent on science, or in other words, scientific research is of fundamental importance for the formation of climate action (Weart 2008; Edwards 2010). Since its start in 1988, the work of the IPCC has increasingly given more priority to assessing the *consequences* of the climate system, due to higher levels of greenhouse gases in the atmosphere, and to *adequate measures*. The IPCC is organized into three working groups that mirror the trinity of: (i) the climate system; (ii) consequences of climate change; and (iii) measures to mitigate climate change. In accordance with a linear knowledge model, these three things are understood as being sequentially ordered. The second and third working groups have gradually been granted greater importance, but the first working group, which consists only of natural scientists, deals with the core of the work and has the highest status, being seen as a prerequisite for the other two.

Since the beginning, the IPCC has been criticized by so-called 'climate deniers', who question what we know about climate change and whether changes, if they exist, are the results of human action. These deniers have pointed to particular uncertainties in the knowledge base and have managed to direct attention to this. Affected by this situation, the IPCC is today putting a lot of effort into increasing clarity about what is known and with what degree of certainty (De Pryck and Hulme 2022). This, in turn, has led to the political issue of climate change being dominated by issues of knowledge and uncertainties. This discussion, to believe or not believe in the results of climate research, as is often heard in the political debate, has for a long time replaced a more genuine political discussion about what we should do about climate change and how society can adapt to a changed climate (Sundqvist et al 2018). Debates over scientific certainty replace political action.

The linear model means that science is independent from politics, though the opposite is not the case. Basically, the model carries a hope that reality will matter, that political elites will attend to important issues that others – politicians and publics – would never have known about if it were not for science. It is a model based on a separation between science and politics which gives priority to the former over the latter. However, as shown in the example of the IPCC, a strong focus on science and scientific certainty risks being a goal in itself.

The deficit model

In relation to the expert, the non-expert is a category that signals a deficit, a lack of knowledge. The general public is often approached as non-expert, and hence as a group of people that lack knowledge and therefore should be improved by education (see Irwin and Wynne 1996 for a critical discussion). Deficits are generally seen as problematic and in need of remedies. This does not mean that the ordinary citizen ought to acquire the same specialized knowledge as the experts. Rather, it is about acquiring enough knowledge to be able to make rational decisions about, for example, health issues and lifestyle choices in relation to climate change. If groups of the public lack the basic knowledge required to keep updated about current societal issues, knowledge deficits can be seen as a democratic problem.

In current modern societies, understanding science and scientific results, on at least a basic level, is a requirement for people's ability to understand developments that may affect them and for them to be able to engage and form a position in relation to scientific applications. Public deficits when it comes to understanding science can also be seen as a political problem that must be remedied, because the public's unintelligible and unjustified critical views can form an obstacle to scientific and technical developments.

The individual's ability to navigate in the expert society as a competent citizen has been discussed by STS scholars in relation to *scientific citizenship* (Irwin 2001). This notion is mainly an analytical tool for analysing the societal distribution of scientific knowledge and differences among groups – not least, the manifold expectations, or lack of expectations, that can be identified in relation to individual citizens. For instance, what is the role of the citizen in an ongoing climate transition? Several studies point to the limited ways transitions include citizens by focusing on *technological fixes*, which allow for only a limited role of citizens as consumers and put trust in experts and technology to 'fix' the climate (cf Barthe et al 2020; Sundqvist 2021). This contrasts with a transparent and public debate, which would allow for a much more varied role of citizens in debates about the direction of an overall transformation of society, about a broad range of solutions and their risks and benefits, and about the involvement of civil society groups, producers, and consumers in downsizing and energy saving.

The *deficit model* (Wynne 1991) expresses a clear boundary between those who have knowledge and those who do not (and thus reveal deficits). In addition, the model is clearly based in the linear model of knowledge, as there is scientific knowledge that can be communicated to the public, who can make more or less sense of this knowledge – that is, show more or less deficit. Both the linear model and the deficit model are one-way models; knowledge travels in one direction only, from scientific experts to the public. It is through informing and educating citizens that knowledge deficits can be repaired. That the public's knowledge deficits are seen as problematic and must be taken care of also shows that scientific issues cannot simply be delegated to experts. The public has an important role to play, either by passively trusting and approving of suggested expert-based political measures or by acting in line with scientific knowledge – for example, by following the advice and prescriptions of medical doctors or becoming more 'climate smart' in everyday life following the instructions of experts.

The high level of interest in the public's level of knowledge can be seen in the regular statistical surveys that are conducted in national and EU populations. The Eurobarometer includes tests of what the population of the

EU knows about various scientific issues, such as synthetic biology, but also what they know about science and technology in general (Eurobarometer 2021). These surveys often ask about simple facts that everyone is expected to learn in school and which can be posed as questions with clear 'true' or 'false' responses (Gaskell et al 2003: 20).

Several of these surveys have been about genetically modified (GM) food, and the factual issues in some questions have been about associations between GM food and contagion, or suggestions that GM food is unnatural or monstrous (Gaskell et al 2003; Gaskell et al 2011). Through such associations, it is suggested that lack of knowledge is also lack of acceptance, in other words: with better knowledge comes a better understanding of the societal use of scientific knowledge, including a friendly and welcoming attitude to scientific and technological development.

There are also examples of interest in the public's knowledge on a more basic level. For instance, in the Swedish Environmental Protection Agency (Naturvårdsverket) survey – reported in *The Public and Climate Change* – the following question was posed: 'Have you heard about climate change?' All respondents answered 'yes' (Naturvårdsverket 2009). This survey also included questions about measures and what respondents know about what they themselves could do for the climate. An understanding of the public from the perspective of its knowledge deficits is based on a clear separation between experts and non-experts and can provide support for one-way information and education campaigns, such as national and European climate campaigns (Uggla 2008).

Beyond technocracy and decisionism

The idea that decision makers should take advantage of available research and expertise implies that knowledge should precede action, which in turn implies that from decisions follows actions and that decisions should be based on knowledge. This assumes a separation between knowledge and action, but also makes an assumption about chronological order: knowledge comes first. For Jürgen Habermas, this order is not, and should not be, the same for all types of question. By distinguishing between questions that can be managed by experts and those that need to be discussed in a public sphere, Habermas finds a way out of Max Weber's pessimistic idea of an ongoing process of rationalization (see Chapter 2).

Habermas' critique of technocracy is based on a distinction between technical problems and political issues. The former can be managed by administration and technical expertise, while the latter requires a political democratic discussion. This distinction is fundamental for Habermas as it enables a critique against scientization – that is, a situation where political issues are treated as if they were technical problems.

In the classic essay 'The scientization of politics and public opinion', Habermas (1971) discusses three models that capture different relations between expert knowledge and politics: *decisionism*, *technocracy*, and *pragmatism*. Decisionism implies a strict distinction between science and politics, as two separate areas of competence and ways of handling issues. According to Habermas, the political sphere can use technical knowledge, but politics always means the exercise of power and the ability to take decisions and prioritize in situations of competing values, and this means that decisions can never be strictly rational (Habermas 1971: 63). Decisionism means that these types of decision are taken by political leaders, without grounding the arguments in public debate. This view is similar to Weber's understanding of politics.

The technocratic model is characterized by expert knowledge dominating the political domain. A far-reaching reliance on expert knowledge means that politics becomes an implementer of measures and strategies that experts suggest. Politics is transformed from being about political will to rational administration (Habermas 1971: 63–64). The room for democratic discussions about what and why, and who is favoured and who is not, diminishes in decision-making processes where citizens have no legitimate role or possibility to contribute to joint understanding and meaning-making. Instrumental rationality takes over when political decisions are no longer about will and choice, but about following imperatives, such as 'according to science', 'technical solutions require', and 'there is no alternative'.

According to Habermas, the existence of both these models in modern societies, despite the crucial difference between them, has strong empirical support. But neither the fact that political decisions are often taken by experts nor the examples of strong political leadership imply that the two models are true in any deeper sense. Habermas argues that a fully monopolized incarnation of political will – that is, assumed by decisionism – does not exist (Habermas 1971: 65–66). The assumption that the technocratic model relies on the idea of a continuous rationality between scientific and political questions is also false. This situation is not a necessity but 'a noteworthy social fact explicable on the basis of objective constellations of interests' (Habermas 1971: 66). The value of the two models is only descriptive. On a deeper level, however, their function is to legitimize undemocratic decisions. Against these two models, Habermas thus sketches a third model.

The pragmatic model assumes a mutual dependency between the priorities of interests and values (politics) and the means to accomplish these priorities (expert knowledge and technologies). The pragmatic model is the only one of the three models that is democratic. Democracy, according to Habermas, is the recognition that values, needs, and interests are shaped in a societal context and in a public dialogue that take advantage of available technical possibilities. Technology will both affect and be affected by these values, needs, and interests. The pragmatic model thus accepts a mutuality between science and politics, but it also assumes fundamental differences between the political and technical spheres.

The pragmatic model is a more constructive alternative to Weber's sociology that can be understood as an impossible choice between further rationalization or charismatic leadership, and which in the interpretation of the Frankfurt School becomes transformed into a dystopia of the Western world's industrialism, capitalism, consumerism, and bureaucratic technocracy.

When Habermas (1984, 1987) developed his theory of communicative action, based on the concept of communicative rationality, he did this in contrast to Weber's concept of practical rationality, which focuses on a means to an end as constituting the theory of rationalization. In a rationalized lifeworld, we are, according to Habermas, attending to the world as if it were tripartite: the objective world, from which we assess whether something is true; the social world, from which we assess whether something is normatively correct; and the subjective world, from which we assess whether a person is honest and means what he/she says. Habermas finds a way that is based on differentiation. The communicative potential in a rationalized lifeworld is about the human being's ability to relate to various validity claims, and it should be seen in light of an understanding of the world as a whole. Thus, differentiation, in this case, does not correspond to an idea of autonomous social spheres. When this communicative potential results in publicly expressed concerns directed to those in government, and aims to make the elite aware of an injustice or other faults for which those in power are held accountable, the question becomes an issue of translating this communicative power to politically binding power (Habermas 1996). For Habermas (1997), differentiation in this sense is an unfinished potential in the modern democratic society.

Political experts

An interesting approach that can deepen our understanding of the relation between scientific knowledge and political decision-making can be found in *controversy studies*. The study of scientific controversies and contradictory knowledge claims is and has always been an important research focus for STS (Pinch 2001; Venturini and Munk 2021). Controversy studies include both pure *scientific controversies* and *mixed controversies*. The latter is of course of most interest to our focus on the interplay between science and politics. These are conflicts around controversial issues that engage additional actors and publics alongside scientists and technical expertise. Mixed controversies are simultaneously scientific and political and develop when debatable issues are mixed with concerns that are not purely scientific. The study of mixed controversies has become a popular field of research that has generated many case studies (Engelhardt and Caplan 1987; Nelkin 1995; Sismondo 2010: Chapter 11).

When there is a controversy, it is easier to explore how knowledge claims and arguments are constituted, what uncertainties look like, and how political judgments enter the debate, compared to when there is no controversy and these things are not debated and made explicit. Not least, it becomes clearer how concerned actors formulate their arguments strategically and in relation to others. Mixed controversies give a vantage point, or provide a peephole, for analysing science–policy relationships.

In mixed controversies, it is important to see the double role of science. Scientific expertise has an important part of the societal production of goods, from food to home electronics and IPCC reports of the climate situation. The development of new chemical products that are widely used today in plastics, textiles, and cosmetics is the result of research. Before these products are distributed in a market, they often go through classification and testing concerning risks to the environment and human health. The regulation of products is managed by other experts – that is, regulatory expertise – who are not connected to industrial production, but to government regulatory bodies. Between these two groups of experts – having the same kind of education – we find tensions and conflicts, concerning the economic interests of industrial production and the focus on environment and health within regulation. This conflict is classical. Of interest in controversy studies is the fact that science, and often the same type of expertise, are located on both sides of the conflict. The available research is the same, but experts have different roles as representatives of *production science* or *impact science* (Schnaiberg 1977; Gould 2015).

Dorothy Nelkin is one of the pioneers of controversy studies, and she focuses especially on what happens when researchers become part of political conflicts. Based on her own conducted research, she formulated the following general thesis about technical expertise and its role in these conflicts: 'when expertise becomes available to both sides of a controversy, it further polarizes conflict. [...] Expertise is reduced to one more weapon in a political arsenal' (Nelkin 1984: 17). This thesis implies that political conflicts shape the assessments made by technical experts. The conflicts are, according to Nelkin, not about scientific facts, but about values that, in the conflict, are put in opposition to each other. The conflicts may be about efficiency versus justice, economy versus risks, government regulation versus individual freedom, or scientific knowledge versus traditional values. These values and their connections to political interests trump expert knowledge.

One important explanation for why controversies arise, and why social values affect scientific assessments, is that scientific facts are fundamentally uncertain (Nelkin 1984: 16). When facts are used in a regulatory context, this often includes legal processes, which are always conflictual and where courts decide which side is right. In these regulatory processes, focus is on uncertainties around the knowledge base. Uncertainties are exposed in a way that they never, or very seldom, are in collegial scientific discussions. At the same time, legal procedures are dramaturgically arranged so that the experts must display as much certainty as possible in order to present themselves as authoritative, and this is also different to collegial discussions in scientific communities.

One example of a controversial process is the application from the Swedish Nuclear Waste Management Company (Svensk Kärnbränslehantering AB – SKB) to construct a final repository for nuclear waste, which was rejected by the Environment Court in January 2018 (Nacka Tingsrätt – Mark- och miljödomstolen 2018; Barthe et al 2020). Legislation regulating this issue has been in place since the late 1970s, with very high demands placed on applications. The applicant must be able to show that there will be no harmful leakage from the disposal to now living or future generations for 100,000 years. To show this scientifically is a task that is alien to established science, but an issue that the nuclear waste company's experts need to manage (Sundqvist 2002, 2012). Management of nuclear waste and assessments of safety issues are controversial. The foundation of this conflict is that scientific facts, irrespective of whether they are about geological conditions, water hydraulics, encapsulation materials, or surveys of public opinion, become mixed with demands posed in legislation and political judgements and in the last resort to the question of whether nuclear power is desirable or not.

In principle, the decision by the Environment Court to not approve the application by the SKB reflected the court's assessment that uncertainties remained concerning the safety of the proposed method for final storage. In January 2022, though the uncertainties pointed out in 2018 still remained, the government decided to approve the application, having assessed the problems as solvable; in this process, the court and the Swedish Radiation Safety Authority (Strålsäkerhetsmyndigheten – SSM) will play a crucial role in defining the conditions for the continued process. Later the same year, the government decided to close down its interdisciplinary Nuclear Waste Council (Kärnavfallsrådet), which had advised the government not to hasten the decision to approve the application. The years leading up to the decisions had included public consultation meetings between the industry, government authorities, environmental organizations, and the public. Seminars and public events organized by the council often brought up controversial issues and counter-expertise, and with its closing, the possibilities for such discussions are narrowed.

At present, there are many questions about what the continued process will look like. How will the industry and government authorities deal with prevailing and emerging uncertainties during the (at least) 70 years it will take before the repository is built and sealed? What transparency will there be in these processes, and what actors and concerns will be involved or taken into consideration? Nelkin's explanation that controversies arise because scientific facts are always uncertain is an argument that can support continuous openness when it comes to processes dealing with long-term safety. The mixed controversy does not seem to have disappeared with the government decision to approve the application.

The results from studies of mixed controversies display a contrast to what many expect from science. The assumption that scientific results can solve political conflicts by showing which side is right is a commonly held view, which is in line with both the linear model and the deficit model. However, Nelkin shows that the situation is exactly the other way around: technical experts are increasing conflicts rather than solving them.

It is possible to identify a fundamental paradox in the science-based conflicts that take place in public as part of debates and political decision-making. Scientific experts who are involved in these processes hold, according to Nelkin, an ambivalent position. They have high status and their knowledge is assessed as the most authoritative that can be found, but they are also often strongly criticized (Nelkin 1975). Their role is about presenting and defending specialized expert knowledge in situations where many other experts with very different competences do the same – for example, in debates on the safe disposal of nuclear waste or the consequences of climate change. There is no single expert that can summarize and clearly present all possible relevant knowledge on such complex issues. In addition, there are always hidden angles to highlight and new questions to pose. Atmospheric chemists are played off against marine scientists, geologists against botanists and social scientists. Experts can strategically attempt to rule each other out when complex issues are discussed, and then the question of which expert to listen to becomes important. There is, however, often a strong ambition to technify issues in political discussions, which means that it is difficult to get away from expert knowledge as such.

Nelkin grounds her studies in a division between *technical knowledge* and *social values*. This reveals a separatist view of the science–policy relationship. However, this dichotomy is nevertheless often negotiated in the controversies studied. Political actors often find it convenient to define issues as technical rather than political. This means that the focus remains on solutions and the efficient means to reach them, which is a way of concealing the values behind both problematizations and suggested solutions (Nelkin 1975: 36).

Nelkin has identified a general pattern in these mixed controversies, which tells that actors initiating something new – for example, when building a new road or introducing a new medical drug to the market – assess issues as technical, while opponents to these initiatives consider the issues as political and value based (Nelkin 1975: 53f, 1984: 18). This pattern is further developed by Nelkin's identification of six tendencies for how scientific knowledge is used in mixed controversies:

> First, developers seek expertise to legitimize their plans and they use their command of technical knowledge to justify their autonomy [...].
>
> Second, while expert advice can help to clarify technical constraints, it is also likely to increase conflict [...].
>
> Third, the extent to which technical advice is accepted depends less on its validity and the competence of the expert, than on the extent to which it reinforces existing political positions [...].
>
> Fourth, those opposing a position need not muster equal evidence [in other words, asking critical questions is often enough] [...].
>
> Fifth, conflict among experts reduces their political impact [...].
>
> Finally, the role of experts appears to be similar regardless of whether they are 'hard' or 'soft' scientists. (1975: 51, 54, emphasis removed)

Nelkin's (1984: 26) own position is that technical knowledge is always uncertain and that a technical framing of an issue tends to hide the underlying political dimensions. She defends a separation between the technical and the political, but focuses on how the two sides interact with each other.

The actors that Nelkin studied acted strategically from their different understandings (a technical or political framing of the issue at stake) to support their own cause. She closely followed these actors in her research to see what they said and what they did. We can conclude that the aim behind Nelkin's studies was to shed light on the role of experts in political decision-making and to make the political dependencies in expert work more visible. In the next chapter, we show how Sheila Jasanoff takes Nelkin's studies one step further along those lines. Before that, we consider Harry Collins and Robert Evans' description of how STS has approached expertise and their suggestion for how we can distinguish between experts and non-experts. This approach implies a demarcation between science and politics, but also a way to defend an understanding of expertise that goes beyond both technocracy and populism.

The third wave of STS

Since the beginning of the 1970s, when STS was an emerging field of research, the ambition has been to explain how scientific knowledge is conditional in relation to social factors. A careful effort was made to distinguish this new research programme from Robert Merton's established *sociology of science*, which focuses on science as a social institution – its norms and patterns of interaction and career paths – as well as how the exchange between science and other institutions works (Merton 1973b). STS was not interested in science as an institution as such, but in explaining how scientific *knowledge* was produced and gained authority. Therefore, the early aim of STS is well captured in the term *sociology of scientific knowledge*, in contrast to Merton's sociology of science. The sociology of scientific knowledge programme aimed to explain scientific knowledge according to social factors that influenced it, which meant distinguishing between the cognitive and the social and explaining the former with the latter.

To explain scientific knowledge with social factors positions these entities as separate from each other but without assuming that scientific knowledge is autonomous of its societal context. Science is part of the social and is a social activity among others. This implies a closeness to Émile Durkheim's classical methodological agenda of explaining social phenomena with reference to social factors (see Durkheim 2014). Scientific knowledge is anchored in the social and can thus be explained by social factors. Based on these assumptions, early STS research formulated a methodological programme which became known as the *strong programme*. This programme was developed by scholars in Edinburgh during the 1970s and has been highly influential in the further development of STS.

The 'strong programme', methodologically developed by David Bloor (1991), includes four principles stipulating that an explanation of scientific knowledge needs to be: *causal* – knowledge is not self-explanatory, but is explained by social factors; *impartial* – analysts should not rely on how the studied actors assess the knowledge in question; *symmetrical* – the same type of explanations should be used for knowledge and beliefs that actors understand as true as well as those understood as false; and *reflexive* – the studies that are conducted in the name of the programme are also in need of explanation and can be studied from the principles of the programme. Many interpreters and followers of the strong programme consider the thesis of symmetry to be the most important, as it can also summarize the whole programme (Collins 1981). Symmetrical studies of scientific knowledge have become synonymous with STS, and discussions about which explanations are more or less, or most, symmetrical have been of great importance for the development of STS and its different subfields (Sismondo 2010: 87, 120–121).

Whether one should make a distinction between the cognitive and the social and explain the former with the latter or whether one should present STS as a sociological programme explaining scientific knowledge with social factors are today under debate and assessed as controversial issues in the field. We return to these and a more thorough presentation of STS approaches in later chapters (see especially Chapter 5), but already here we introduce an approach, formulated by leading researchers within the field, with the aim of understanding the relation between science and democracy. This approach makes explicit connections to the roots of STS and also makes a sharper distinction between science and politics than the approaches we meet in subsequent chapters.

In 2002, Harry Collins, who has been one of the leading researchers in STS since its beginnings, published an article together with Robert Evans in which they characterize the development of STS according to three waves (Collins and Evans 2002). In the first wave, which came before the establishment of STS as a research field, science was considered autonomous and value free, and therefore was seen as being capable of solving political conflicts. During the 1950s and 1960s, this view dominated both academic research and public debate. This first wave is what pioneers in the STS field criticized, and it is from the standpoint of this critique that the field then developed.

The second wave is characterized by an understanding of science as a social activity and was, thus, dominated by a social constructivist perspective. According to Collins and Evans and from the principles of the strong programme, from this perspective, explanations of scientific development – for example, how scientific controversies are solved and how some knowledge is perceived as true and other knowledge as false – social factors outside science – such as political, economic, or religious interests – must play a role.

The second wave has been successful and has made STS into an established research field. However, according to Collins and Evans, during the development of the second wave, the reaction towards the first wave was taken too far. Knowledge and expertise are no longer recognized as something *substantive*, but as an attributed quality. In other words, an expert is defined as someone who is labelled 'expert' (Collins and Evans 2002: 239), and thereby the content of knowledge disappears.

Collins and Evans have continued to develop the third wave since 2002. This approach assumes that scientific expertise is real, which means that STS scholars should be able to tell what expertise is and also what is good and less good knowledge. This, they argue, is about defending the vertical dimension of epistemology (Collins and Evans 2002: 268). The consequence of not taking a stance in relation to who possesses more or less knowledge, they argue, is to advocate a relativistic position which runs the danger of turning into knowledge populism.

> Like it or not, those who study knowledge are experts in the nature of knowledge. If we refuse to acknowledge any role other than criticism – if we are willing only to level down and never to build, explain, or evaluate the structure of the vertical dimension of epistemology – we are evading a responsibility that only we can fulfil. (Collins and Evans 2007: 140)

Collins and Evans (2007: 134) suggest that it is the task of the STS researcher to answer the question of how knowledge should best be assessed – as good, relevant, and so on – in a specific decision concerning science. This task, however, requires that a distinction is made between who is an expert and who is not. To refrain from this, they argue, is equal to defending a society that has turned into a nightmare (Collins et al 2010).

Collins and Evans take many insights from the second wave into their third wave, but their approach is different from what they describe as social constructivism, in which all knowledge is made equal. The third wave understands knowledge as real and is able to see the differences between what is knowledge and what is not, and who possesses knowledge and who does not. In this way, the vertical dimension of knowledge is restored. The important critique of technocracy, as based in autonomous science, has gone too far and risks ending up as a populist position in which there are no boundaries at all between different kinds of knowledge and actors.

According to Collins and Evans, the analysts should not just limit themselves to the study of other actors' boundary-making; they also need to draw their own boundaries. Wave two brought the important insight that scientific knowledge is social, and thus not autonomous. But this does not mean that all activities and competences are equal. Being a researcher and a scientific expert implies a social activity among other social activities, but one based on knowledge and competences that some possess and others lack. Competences and skills are not evenly distributed.

The third wave does not understand scientific knowledge as something out there and socially independent. Going back to the first wave is not an option. The ambition is to develop the sociological understanding of scientific knowledge by anchoring it in an understanding of society that is real and not merely about how actors assess knowledge, without discriminating between different assessments. Our view, however, is that this critique of the second wave is exaggerated. In wave two, Collins and Evans lump together quite different approaches, and some are more relativistic than others. For instance, the strong programme developed by Bloor does consider scientific knowledge as a real social phenomenon, something to be explained. If this was just an attribution by a specific actor, there would not be much to explain. The two Edinburgh scholars Barry Barnes and David Edge, co-developers of the strong programme, summarized the situation so far in 1982 when they claimed that expertise is something social but also something *special* – the source of cognitive authority (see Barnes and Edge 1982: 2–3).

STS scholars, who are specialists on scientific knowledge, should, according to Collins and Evans, be able to sort between different kinds of knowledge. This is the most important component of being specialist on knowledge – being able to assess what knowledge is and how it works, what it can do and what it cannot do. Part of this ambition is to give advice in discussions about what social problems scientific knowledge can help to solve.

Gil Eyal (2013: 870–871) makes an important addition to the discussion about the third wave. Eyal agrees with Collins and Evans that expertise is a substantive skill possessed by individuals and obtained by socialization within groups of specialists. However, he disagrees that STS scholars can or should make decisions about what expert possesses the best knowledge. A substantive view on expertise is important, but this should be part of a descriptive agenda studying skills and competences – the distribution of expertise – without deciding what expertise is correct. To sort out who is the most competent expert would lead to controversies about the expert knowledge of the individual STS expert. According to Eyal, this aim is not a necessary part of a substantive view on expertise, and it is much more complicated than Collins and Evans assume.

Scientific knowledge is important for solving the problems defined as being of a technical character. But for scientific knowledge to be useful for decision-making, researchers need to agree. When there are scientific controversies, Collins and Evans argue, the vertical dimension of epistemology collapses, and we then need to trust our democratic institutions and let the political system decide without being affected by conflicting experts. In other words, the stronger the scientific consensus, the stronger the compelling force that politicians should feel to act in accordance with scientific knowledge (Durant 2016: 18).

Similarities can be identified between the third wave of STS, Haas' ideas about the importance of scientific consensus, and the linear model of knowledge. Issues for which there is relevant knowledge and a scientific consensus can be treated as technical issues. However, Collins and Evans agree with most STS scholars that technocracy is a real problem in the modern science-dependent society; though when criticizing the second wave in STS for taking the critique against technocracy too far, they raise the opposite problem of populism and argue that STS scholars risk supporting a populist position by being reluctant to take a stand on what knowledge is and who possesses it. The political counter-reactions against technocracy are what Weber terms 'charismatic authority' (Weber 1983: 166), which is based on expectations that political leaders carry extraordinary abilities in exercising an autonomous political will. A current example can be seen in former US president Donald Trump, and this is what Collins and Evans term 'populism' (Collins et al 2020). According to Collins and Evans, there are better alternatives to technocracy than charismatic leaders, and STS has an important role to help advise on this alternative. A start would be to better understand what expertise is.

Conclusions

In this chapter, we have met separatist views on science and politics, but in contrast to Chapter 2, these have not focused on the autonomy of the two fields, but on how they relate to each other. Since this is a book about STS, it is important to note that we find a clear separatist ambition in early STS, exemplified by the strong programme and Nelkin's studies of controversies. In these works, a clear distinction is made between scientific knowledge and the social (political) context. The former is a social but cognitive entity that could be explained by other social factors. In our view, this early STS position is taken up in Collins and Evans' third wave, but the critique is not aimed at the strong programme but at those STS scholars who, in the name of STS, have taken the methodological principle of symmetry too far by interpreting this as a tool for making all knowledge under study of equal quality. However, according to Collins and Evans, all knowledge is not equal. The principle of symmetry stands on the firm sociological foundation that *if* scientific knowledge has gained authority and high status, this fact – the acquired authority and status – needs to be explained by social factors. This analysis will reveal power inequalities as well as social interests, but does not change the fact that science is still rewarded authority and status – that is, something real and something special.

Symmetrical analyses have focused on particular cases, showing disruptions and paradigm shifts as well as how science has been used for particular purposes. STS analyses show that this is a matter of *knowledge relevance* rather than a question of truths or facts that stand alone before their relevance has been established, yet this enables us to ask: For whom is this relevant, and for what purposes?

After chapters 2 and 3, we can conclude that separatist views are common. They have followed closely in the footsteps of modernity and can be traced back to ancient Greeks, such as Plato, and we have identified and assessed them as crucial in early STS and in current STS, for example in Collins and Evans' third wave approach. However, it is also important to emphasize that STS scholars do not adhere to separatist views that universalize the separation between science and politics and understand them as ontologically different and universally stable.

The STS position is about being open to change, including the shifting boundaries and dynamic interplay between science and politics. We see many examples of this in following chapters. An important contribution by STS scholars is provided by the many empirically oriented studies on how actors try to separate and quite often defend their positions – in conflicts with other scientists as well as with other societal actors in so-called mixed controversies – by trying to *universalize* them – for instance, when claiming that safe disposal of nuclear waste is only about 'certain knowledge' and that this is what science can provide. Such unrealistic arguments fuel the kinds of controversies studied by Nelkin and are also an important background for the STS aim of presenting a realistic understanding of how science works in society, though as we have seen, this 'work' *is* often made up by unrealistic positions and ambitions. This ambition is critical of technocratic strategies and could act as a kind of vaccination against scientific hubris (Jasanoff 2003). The STS agenda means to contribute to a better understanding – a more realistic, nuanced, and truer picture – of how science works in society and what scientific knowledge can do and cannot do.

One move towards a more realistic picture of science, which STS scholars have contributed to, is the conclusion that scientific knowledge is never autonomous and that the linear model and the deficit model are not accurate in terms of how science and scientific knowledge work. Yet, both the deficit model and the linear model are common strategic weapons for all those who want scientific knowledge to guide political decision-making and aim to highlight science as what is most important in society. As such, they are *real* in their consequences and need to be given serious consideration by STS analysts who want to understand how science works in society. Thus, STS provides an empirically grounded *shadow theory* of modern enlightenment science by showing how it really works, and critically studying the separation between science and democracy while simultaneously acknowledging its practical importance is part of this ambition.

PART II

Overlap

4

Close but Not Too Close

Introduction

As an interlude, this chapter is an effort to connect, bridge, and provide an overlap between the other two parts of this book: Separation and Co-production. In this work, we complicate the picture of a clear division between science and politics. Our focus is on science and politics as two separate activities with their own institutions and practices, but we also turn to the question of how the interplay between science and politics is understood in terms of overlaps and connections.

Better communication, increased proximity, or more interaction while maintaining an arm's-length distance – there are many expressions of how science and politics connect and should be connected; and from the science side, they can be summarized as an ambition to make science relevant to policy, to make it close to politics but not too close (Gieryn 1995: 435, referring to Jasanoff 1990). In other words, policy-relevant research involves a balancing act between separation and integration (Sundqvist et al 2015, 2018). According to this view, research must be sufficiently separate to maintain its autonomy, but integrated enough to be socially relevant. This ambition entails a critique of an overly strong division, but also points out the risks that one side will take command of the other. From this perspective, a scientization of politics and a politicization of science are both highly undesirable and considered as risks. There is thus a need for a proper distance between science and politics.

To be fair, we must say that several of the approaches and perspectives presented in Chapter 3 devote a great deal of effort to understanding the interplay between science and politics. Collins and Evans' third wave suggests such a 'balanced' solution. This also applies to Habermas' pragmatism as well as to Nelkin's aim of understanding scientific experts as part of a political context and, from there, sorting out the conditions for expert work.

However, the aim of chapters 2 and 3 was to account for a division between science and politics and the approaches that focus on this, including suggestions from some scholars that one side should dominate the other. Here, Plato's allegory of the cave is archetypal, as it is based on a strong division but also on a desired dominance of knowledge over politics. This allegory is echoed in delegative democracy, the linear model of knowledge, and the deficit model. In these approaches, a clear division is expressed, but also a desire that expert knowledge should influence, or even dominate, publics' everyday knowledge.

In this chapter, we present approaches that accept a division between science and politics but devote most of their efforts to thinking about the exchange and the desired balance between the two and, not least, about clever combinations through greater proximity, better bridges, and better mixes. We encounter approaches that have an explicit policy ambition for how science and politics should interact. We also give examples of current ways of organizing the interplay between science and decision-making and of organizations that have an ambition to do both science and politics. Therefore, one central issue in this chapter is boundaries in the space between science and politics, and how both individuals and organizations can operate within these boundaries. We focus on how individuals, groups, and organizations manage their work as scientific experts and boundary organizations. In a final section, we introduce the idea of hybridization in relation to some key developments around science, technology, and innovation that question the boundaries, distances, and linearities we present.

Research policy and the social contract between state and research

From the middle of the 20th century, scientific knowledge has become increasingly important for political decision-making. The development of the industrial welfare state implies that politicians seek the help of scientific experts to justify public spending by referring to expert knowledge. Major investments in research have also taken on great importance in nation states' attempts to assert themselves in global competition (Jasanoff 2017: 259ff). Therefore, scientific experts have become increasingly important as advisors in political decision-making. This pattern is evident throughout the Western world, and during the Cold War, the same logic could be seen on both sides of the iron curtain. Conflict and competition were directed by scientific development in the arms race and the space programmes. In addition, in the US, social scientists started to study how science is organized within the framework of large programmes, guided by the wishes of governments to achieve political, economic, and military progress; and how researchers develop the role of scientific advisor. Research policy studies became a research specialty in its own right (Fagerberg et al 2012).

The US nuclear weapons programme aimed to produce an atomic bomb, the so-called Manhattan Project (1942–1945), is often seen as a paradigmatic example of the shaping of the interplay between the state and the scientific community. The 'social contract' developed between the parties entails clear state governance, but a governance that at the same time safeguards the independence of research (Ziman 1994; Frank and Meyer 2020). In all democracies, freedom of research is an important component. Research dictated by political power is a clear feature of authoritarian regimes. A dictatorship does not leave research free, although the conditions of research in a democratic as well as in an authoritarian state can show great variation (Salomon 2000).

To be controlled but at the same time be independent sounds like a paradox. The solution proposed by the social contract is partial freedom. Political power governs by providing tasks and resources to solve the tasks, but research has the freedom to solve them in the way it finds is best. This interplay has also resulted in what has been called 'mandated science' (Salter 1988). Many follow-up issues arise and must be addressed in the concrete formulation of such a research policy. What research should the state invest in based on society's needs? In which research areas can scientific results be expected, and in which areas is it less meaningful to invest resources? How should targeted investments be organized: into large programmes or several smaller ones? How should the state control the freedom of research and offer a reasonable level of freedom? And how can scientists guard their own (relative) freedom?

The controllability of research has long been discussed, and an important reference is Michael Polanyi's classical text from 1962 in which he proclaims 'the republic of science'. Polanyi argues that the results of science are unpredictable, and this applies even more to the usefulness of the results. Management of research based on an imagined benefit will therefore be a rather unprofitable project. Furthermore, Polanyi claims that scientific work is a craft with large elements of tacit knowledge, which makes competence difficult to transfer and communicate to those who are not themselves part of the specialist culture, and thus difficult for outsiders to control. Researchers always know more than they can tell (Polanyi 1966). All in all, this means that research is by nature and necessity a free activity (an independent republic), even if this is not considered desirable.

In another classical research policy study from the same period, Don K. Price (1965) argues for an almost opposite position, namely that researchers and experts in the early 1960s became the most important political establishment in the US. He supports President Eisenhower's warning that science has become increasingly linked to military and industry interests and that scientists have become a new priesthood, far from the role of representing an enlightenment that leads humanity away from tyranny. Freedom seems to disappear and, at the same time, democratic governance of research needs to be vigilant because researchers have become a threat to democracy. However, Price also had ideas on how knowledge and power could be better balanced with the help of policy-trained experts (see Jasanoff 2017: 264). Questions about who controls whom and with what results have, since these classical texts, continued to be under lively debate.

The main reason for the development of state-led research policy is the competitive situation between nation states, from both a military and an economic perspective, and in particular between the Great Powers. It is therefore no coincidence that the Manhattan Project has assumed iconic status. The space programmes during the Cold War became a logical continuation of the development of nuclear weapons. The military arms programmes were followed by civilian programmes, such as nuclear technology programmes and environmental monitoring programmes. Early on, environmental research had its major research programmes strongly influenced by a national competitive situation. The Swedish soil scientist Svante Odén, who was the first to describe regional acidification in Europe, spoke of chemical warfare between nations based on the idea that emissions in one country create environmental problems in another (Odén 1967; Lundgren 1998). Climate research is a continuation of this pattern as it shows that greenhouse gas emissions at a specific location affect all humanity. Climate change implies chemical warfare at its global peak. However, environmental research and not least atmospheric research, even if born out of the arms race, has been characterized by strong international cooperation (Edwards 2010, 2012).

It is not only science and technology that are governed by an active research policy, and it is not only for military and economic reasons that investments in research are made. In the expansion of the welfare state, social science research has been systematically used by political and reformist agendas in developing the welfare state in terms of medical care, education, gender equality, housing, and urban planning. However, it is not only states that pursue an organized research policy. In today's global competitive situation, most private companies invest in and use research to strengthen their positions. Research is useful to many, and in several different ways (Nowotny et al 2001).

To be effective, the delivery of resources to researchers, called 'research policy' requires competent clients. The delivery of knowledge from researchers to the surrounding society, or 'science advice' – that is, a transfer that goes in the opposite direction to research policy – is also a major challenge. It is not often that researchers can be content with submitting the latest articles published in specialized scientific journals to policy makers. Extensive translation work is required to summarize research in such a way that it becomes understandable and useful for those who are not specialists themselves but may be able to benefit from the results of research.

The linear knowledge model (see Chapter 3) pays close attention to the chain of translations from basic research to commercial products and public goods. From this perspective, making knowledge useful means adaptation and hard work from both sender and receiver so that a fruitful exchange can take place along the chain which will satisfy both parties. This can be expressed as the sharp edges of the autonomous spheres – science and politics – having to become softer for fruitful meetings to take place. However, a less linear and more interactive model is also interested in how different kinds of developers and users interact in all the phases of the development and spread of knowledge.

From science to expertise and advice

In this book, we have talked about science but also about experts, including scientific experts, and expertise. To clarify the use of terms, it is appropriate to set out the difference between researchers and experts, and between science and science advice. This distinction is based on the practical activities involved. The first activity is performed in the laboratory by researchers, who, in a specialized environment together with colleagues, produce specialist knowledge. In this world, colleagues – those who have access to the same type of specialist knowledge – are co-workers and critical reviewers as well as recipients and part of the audience. This suggests a limited and highly specialized collegial world. But the specialists are not specialists for their own pleasure. There are expectations that the produced specialist knowledge can lead to knowledge, applications, and products that are useful to others. For this reason, researchers and specialists are given resources to develop their knowledge. The state, industry, foundations, or interest groups pay for the work of specialists and hope for a useable return. This can be seen as an exchange, or a contract, between researchers and funders, where specialists are given resources to make research based on promises to deliver useful results. This is what delegative democracy tells us and what research policy is about.

When researchers present their results to the outside society, they become experts and scientific advisers. As such, they are expected to bridge the gap between science and politics. The task of giving science advice is about summarizing the state of scientific knowledge – that is, referring to what science says – but also about making knowledge relevant (Barnes and Edge 1982: 233–237). To become experts, researchers must leave their white coats and the protective scientific environment in the laboratory and begin to interact with non-specialists. Researchers become scientific experts. A scientific expert is a person with specialist knowledge who summarizes this to make it comprehensible and useful to others who are outside the world of research. In other words, scientific experts are messengers and mediators – that is, science advisors (Sundqvist 2000).

From Chapter 2, we remember that Plato saw the messenger between the world of knowledge and the social world as experiencing great difficulties. Few are privileged to be chosen to carry out this mission, and in the execution the messenger is exposed to both the glare of light and the distrust of cave dwellers. Latour believes, however, that the situation in today's knowledge societies does not have many similarities with Plato's allegory. Instead of a cave opening, there is today a wide boulevard where researchers can move in a large space between 'the social world' and 'the world of ideas' (Latour 2004: 11). The passage is no longer narrow, and it may never have been. But even though mediators of knowledge are many in a modern society, efforts are still required in order to create useful knowledge.

Useful knowledge

During the post-war period, American political scientists interested in policy analysis focused on the interplay between science and politics, and not least on the question of what knowledge is needed for good political decisions (Jasanoff 2017: 263). As mentioned earlier, the US government has also systematically worked to organize science advice in an effective way and has created an activity – research policy – which social scientists have made into a study object. Although science advice is organized as part of the social contract where funders – such as governments – influence what researchers do, it is often carried out according to the aim that scientific knowledge is independent of political judgments and decision-making. It was in this tradition that the expression 'speaking truth to power' was born.

Raphael Sassower (2014: 2) claims that 'speaking truth to power' was coined by the Quaker society in the US in 1955 in relation to supporting nuclear disarmament. This means that something needs to be done (it is time for action) and that there is important knowledge to consider for those in charge. Important questions arise concerning who is the sender (who has access to the 'truth') and who is the receiver, and not least how the contact between sender and receiver can and should be established. Scientific knowledge is not useful in itself; for scientific knowledge to become 'useful knowledge', translation work is needed. The knowledge needs to be packaged and repackaged.

The American research tradition on science advice assumes that connecting science and politics is problematic precisely because it is about a meeting between autonomous spheres. Robert Merton (1945) talks about frustrations emerging during this exchange. There is also the idea that science must adapt to politics. Frictions and frustrations are not just the fault of politicians (ignorant decision makers); scientists must also contribute in order for a positive exchange to take place. As early as 1959, Charles Lindblom warned of the problematic nature of abstract science as a basis for political advice, arguing that theoretical scientific knowledge is of limited value to politicians and bureaucrats. Useful knowledge necessarily includes everyday knowledge, which needs to be based in a problem in need of treatment (Lindblom 1959: 81).

Lindblom's research is based in pragmatism, which implies moving from a linear model to a more interactive model. The exchange between science and politics is understood as a communication problem where both sides must adapt and contribute to an effective use of research if good outcomes are to be achieved. Both autonomy and a one-way (linear) exchange – which is based on the dominance of knowledge and focuses on the lack of competence and interest among receivers while taking the knowledge of the sender for granted – are questioned.

A common way of understanding the relationship between science and politics is to distinguish between the supply (push) and the demand (pull) sides of scientific knowledge (Sarewitz and Pielke 2007). If we place the two dimensions of supply and demand in a matrix, we get four fields that generate different answers to the following two questions: 'Is relevant knowledge produced?' (supply side) and 'Can users assimilate relevant knowledge?' (demand side) (see Figure 4.1). According to Daniel Sarewitz and Roger Pielke (2007: 14), we can find many examples of poor matches between supply and demand which reflect that users are unable to assimilate relevant and accessible information, but also that science fails to produce relevant and useful knowledge. If both the questions in Figure 4.1 are answered with a 'no', we have a genuine separation problem. According to the proponents of the matrix, both producers and consumers have a responsibility to translate knowledge into something that becomes useful. Avoiding strong separation is the goal when assessing activities in relation to this model, and the field at the bottom left of the matrix is the ideal.

In many of the studies on how the interaction between science and policy works, it is the recipients (politicians) who are problematized. This is where the linear model and the deficit model come to the fore – that is, experts are always right, and if politicians and the public do not understand, it is they who must change (they must be educated or educate themselves). The consequences are in focus here – how the message (knowledge) is understood and used and what happens as a result of this. But we also find several studies, especially in STS research, questioning scientific experts by focusing on their shortcomings (Irwin and Wynne 1996).

Nico Stehr and Reiner Grundmann (2012: 34) believe that different actors and their opportunities to act must be better linked for knowledge to become 'practical knowledge'. It is increasingly argued that science must become better at packaging knowledge before it is presented to decision makers, so as to make it useful and not just potentially useful.

Figure 4.1: The supply and demand of knowledge

		Demand: can users assimilate relevant knowledge?	
		YES	NO
Supply: is relevant knowledge produced?	NO	The research agenda is inappropriate.	The research agenda and user needs are poorly matched, and users may be disenfranchised.
	YES	Empowered users take advantage of well-deployed research capabilities.	Unsophisticated or marginalized users, institutional constraints, or other obstacles prevent information use.

Source: Sarewitz and Pielke (2007: 12)

Peter Haas, who we met in Chapter 3 as a defender of the autonomy of science, has developed a three-step model for how the autonomy of knowledge should be balanced in relation to usability (Haas and Stevens 2011). In a first phase, when the agenda is formulated, it is important that knowledge actors – scientific experts – are autonomous. In the second phase, when knowledge is formulated and established, it is also important for research to have autonomy. If the researchers in this phase reach agreement, the process can move on to a third phase. In this phase, when knowledge is to be communicated to decision makers, it is important that experts become as connected as possible to politics. In short: during the first and second phases, distance is important, while in the third, proximity should be a priority. Autonomy has its place, but so do close interaction and collaboration.

Scientific experts

Science advice is a specific part of research policy. It is not about how the state controls research, but how researchers and research results can have an impact on the state and other decision makers. Science advice is about how knowledge is used in the management of social problems, regardless of how this knowledge has been controlled and organized based on research policy purposes. Researchers have something they want to point out to decision makers, whether this is desired or unwanted by the receivers. Speaking truth to power is an example of science advice that is not demanded by policy makers, and therefore often undesirable. In environmental research, we find many examples of this relationship, such as the discovery of a depleted ozone layer that came to be understood as an ozone hole. This case, however, ended with political leaders listening to scientists and acting decisively together with industry (Litfin 1995). Usually, however, science advice is about commissioned research that is communicated to decision makers in accordance with the social contract where researchers are guided by decision makers' wishes but at the same time exercise partial freedom. The researchers' exercise of freedom then leads to products that are delivered to decision makers, which may involve anything from nuclear weapons to assessment reports on the climate system.

Findings of studies carried out by climate researchers were initially not demanded by decision makers, but since the establishment of the UN Intergovernmental Panel on Climate Change (IPCC) in 1988, reports on the findings of these studies have been commissioned (Weart 2008). The governments of the UN member states are responsible for initiating the processes that led to the production of the IPCC reports, which means being responsible for the timetable and structure of the reports as well as deciding who the authors will be. The selected researchers then independently summarize the research based on the structure that has been decided. The climate research found in the reports has already been published in scientific journals but is summarized by the IPCC, based on the idea that this work is a necessary part of performing science advice. Knowledge does not speak for itself; it must be organized and summarized to become useful. The organization of the IPCC can be seen as a contract between the UN and the global research community (De Pryck and Hulme 2022).

A comprehensive review of research on science advice, compiled by the European Commission's body Science Advice for Policy by European Academies (SAPEA), argues that the issue of science advice is more important than ever (SAPEA 2019). The need for science advice is urgent for the handling of today's major societal challenges. Human interventions and the impact they have on nature and society are greater than ever, and the climate crisis is a clear example. However, at the same time as problems are growing, knowledge has become more complex. In the report, this is summarized in the following way: where knowledge is mostly needed, it is most complex, interdisciplinary, and uncertain, and this unfortunately also applies to the knowledge of science advice. In today's context of populism, polarization, and fake news, science advice is of fundamental importance for the preservation and development of a democratic society, it is argued. Evidence-informed decision-making is advocated by SAPEA, which does not mean that decision makers should be guided by experts, but that they should listen to them and carefully evaluate their advice.

SAPEA is a scientific advisory body that advises EU commissioners to improve their decision-making. SAPEA represents more than 100 academies from more than 40 countries in Europe and is one part of the European Commission's Scientific Advice Mechanism. The second part consists of the European Commission's Group of Chief Scientific Advisors, which consists of up to seven leading researchers. The reason for setting up this mechanism was the Commission's view that science advice is crucial for EU decision-making. It is thus no coincidence that SAPEA produces a compilation of knowledge which concludes that science advice is a central and urgent issue that has never been more important than today.

The SAPEA report is an ambitious summary of research on science advice, and it also formulates clear advice about what can be learned from this research. The starting point is that science advice constitutes a complex landscape where concrete advice can take many different forms in terms of both content and organization. The importance of avoiding scientific, rational, and linear models for understanding and organizing science advice is emphasized. The advice should be inclusive in terms of both knowledge and actors, and it is emphasized that more than scientific knowledge is needed for a successful result. Science advice is not a strictly scientific activity; it is value based and linked to decision-making processes, which justifies the inclusion of interest groups and the public.

Despite its broad perspective on science advice, the report takes as its starting point the scientific side of the interplay between advice and decision-making. The primary focus is to get the scientific knowledge right for presentation to decision makers. And the 'right way' means that the scientific advisors should attach great importance to methodological rigour, quality assurance, and peer review to communicate the best available knowledge in a transparent way, without being prescriptive. The emphasis is on how information is to be given, and less interest is devoted to decision-making processes and the problems that are to be solved in those. Involvement of the public and types of knowledge other than science-based ones are often forgotten when advice is understood from the science side. In summary, the report is based on a sharp distinction between scientific knowledge and decision-making processes, and the question that is asked is how the former can best inform the latter. The view is clearly separatist and based on a linear view, even though the linear approach is explicitly recognized as too narrow and limited. This shows how deep the linear knowledge model goes.

The separation between knowledge on the one hand and problems and decision-making on the other becomes even clearer in the report from the Group of Chief Scientific Advisors (2019), which is a subsequent comment on the report from SAPEA. Here the scientific side is emphasized even more strongly. According to the group, well-functioning science advice means that high-quality research must be placed at the centre and that researchers who represent this research must act as mediators between science and policy. The implication is that if only the best scientific knowledge is conveyed correctly, good decisions will be made.

The research policy landscape

In a classical study *Science Speaks to Power: The Role of Experts in Policymaking*, David Collingridge and Colin Reeve (1986) argue that the interplay between science and politics oscillates between under-critical and over-critical situations, both of which inhibit the impact of knowledge on politics. An insoluble and almost tragic situation is presented in which the two worlds never meet. In an under-critical situation, politicians have agreed on the agenda and research results are used to confirm politicians' views on the matter. In an over-critical situation, the political dividing lines are sharp and expert knowledge is of interest to decision makers only because of expectations to resolve political conflicts. But science cannot live up to such expectations. What happens is that scientific arguments contribute to prolonging and deepening political conflicts. This brings us back to the under-critical situation, but now it is in the form of conflict between different camps that confirm their political positions with the help of scientific arguments.

Collingridge and Reeve's analysis is based on the differences between science and politics. According to these, successful scientific work has three characteristics: autonomy, disciplinarity, and consensus (being devoid of criticism threatening this consensus). This characterization is based on Thomas Kuhn's (1962) concept of 'normal science'. Politics, on the other hand, is about making compromises between different and sometimes incompatible interests, and good politics is characterized by openness to criticism. When scientific knowledge is used as a basis for political decision-making, the characteristics of both science and politics are lost; scientific facts are negotiated and political differences of opinion as well as criticisms are obscured by scientific facts. In addition, what Collingridge and Reeve call 'error costs' increases dramatically when knowledge moves from science to politics. In the scientific laboratory, errors are cheap, unlike when scientific knowledge is used in political decision-making, where human lives and large financial sums are often at stake.

When mistakes are costly, as they are when it comes to climate action, the degree of criticism increases, leading decision makers to seek more and better scientific facts. But since science does not produce unambiguous facts (at least not at this complex level), there will always be experts as well as politicians who criticize both the decisions and the facts on which these are based. We get an endless debate about what is correct and safe and who can be trusted among researchers and decision makers as well as other actors. The discussion moves further and further away from a democratic debate on clear political alternatives. Collingridge and Reeve give a pessimistic view of an existing separation that cannot be bridged, but where attempts to do precisely that dominate, leading to practical failures and broken expectations.

However, the space between science and politics is more interesting than Collingridge and Reeve have understood (see, for instance, Nelkin 1984; Jasanoff 1990; Sarewitz 2004). Science is not a monolith; a diversity of scientific disciplines and research areas exist, and these are also changing. New interdisciplinary fields arise in response to societal problems. Moreover, science advice is not about presenting 'laboratory knowledge' to decision makers. Useful knowledge is not to be found on the science bookshelf, but is created in interaction between researchers and the actors who are interested in using the knowledge at stake. A sharp line between science and politics is neither reasonable nor desirable. Such views have been taken further in post-normal science and Mode 2 research. Like Collingridge and Reeve's approach, these two approaches are based on an understanding of a fundamental difference (separation) between science and politics, but they try to overcome the deep pessimism we find in the 'critical' models.

Silvio Funtowicz and Jerome Ravetz (1993), who coined the term 'post-normal science', are interested in issues that are so complex that they cannot be handled by scientific disciplines. When dealing with such issues, boundaries between disciplines must be resolved, as must the boundary between scientific and non-scientific issues. These issues also include Collingridge and Reeve's increased error costs and what Funtowicz and Ravetz (1993) call 'decision stakes'. The issues that are in focus for post-normal science are characterized by knowledge uncertainties and the fact that great societal values are at stake while, at the same time, there is political pressure to act on the problems. With the concept of post-normal science, Funtowicz and Ravetz try to solve the problem that Collingridge and Reeve present as an insoluble social tragedy.

When Kuhn's normal science – in which uncertainties are handled internally and almost automatically and where values are unspoken – no longer works, a new science emerges, which is characterized by both uncertainties and values being made explicit and consequently requiring conscious management. This is the post-normal situation, which means a scientific practice that is reflective, inclusive, and transparent regarding both scientific uncertainties and decision alternatives. Science cannot unambiguously solve complex societal problems, but it can become better at supporting solutions by openly reporting what science can contribute. The approach of post-normal science is hopeful. Science can adapt to a complicated situation and contribute to dealing with complex societal issues.

The description of a post-normal science has had a great impact on academic discussions and resonates with another popular typology for understanding the situation of research in a new research policy landscape. Helga Nowotny and colleagues describe an ongoing change in research practice, which they believe goes from autonomy to collaboration; this is viewed as the development from Mode 1 to Mode 2 research (Nowotny et al 2001).

Here too the starting point is Kuhn's concept of normal science, which is called Mode 1. In this situation, the disciplinary context is important and the scientific activity is clearly specialized and separate from the rest of society. In contrast stands Mode 2, which implies that research is developed in an applied and multidisciplinary context and with close collaboration between academia, government, and business. Stakeholders are invited early in the research process to discuss the possible usefulness of the results. The autonomy of science and the sharp boundaries between disciplines and the outside society are dissolved, and this is done in a deliberate way. Research is initiated and developed in the context of application. The purpose of research is to solve societal problems, and this is often a matter of strengthening economic competitiveness. The focus is technological, and the key word is 'innovation'.

In Mode 2, knowledge production is spread over several different actors and exposed to external pressure in terms of focus and results. Knowledge must be useful. Nowotny and co-authors claim that knowledge production according to Mode 2 opens for a social robust knowledge. It is then a matter of toning down science's claim to objective and universally valid knowledge. Broader participation and recognition of more and different requirements than those important in academic and disciplinary research are central. Usability, which is understood as social robustness, replaces truth as a decisive knowledge criterion.

The context for knowledge production is radically different in Mode 1 and Mode 2, but this does not mean that they are mutually exclusive. In the concrete production of knowledge, as well as in science advice activities, there has long been a mixture of the two forms (Martin and Etzkowitz 2000). Despite tendencies towards dissolved boundaries and more collaboration, science is still an exclusive institution. The fears raised in connection with debates about post-truth show that there are many who defend the traditional organization of science – Mode 1 – as a reaction to the fact that truth is reduced to usefulness and benefit. However, the ideas of post-normal research and Mode 2 research show that science is adaptable and that much is going on in the space between science and politics.

Since Nowotny and her colleagues described the differences between Mode 1 and Mode 2, developments have occurred that could be seen as an escalation of some of the relations pictured in Mode 2. Elias Carayannis and David Campbell (2012: 3) describe a knowledge production system that is 'reinforcing innovation networks and knowledge clusters consisting of human and intellectual capital, shaped by social capital and underpinned by financial capital'. They even go as far as to call these developments Mode 3.

'Open science and innovation' is today a policy priority for the European Commission. This has implications for how citizens are expected to engage with science as well as wider science–society relations. For example, the European Commission (2020) expresses that through open science and innovation, including partners from across academia, industry, public authorities, and citizen groups, both the creativity of and trust in science increases. Also this initiative could be described as an escalation of a Mode 2 situation. In critical studies of open innovation and science, this 'new age of openness' has been described as being revolutionized and facilitated by digital infrastructure and situated in a market economy where science and innovation are increasingly under pressure of commercialization and popularization (Smart et al 2019; see also Mirowski 2018). In this new situation for knowledge production, citizens are approached more frequently as users and co-designers of open innovation and are thereby given a narrower role than when approached as concerned publics, as collectives and citizens.

Rather than pointing to definite breaks between different modes of knowledge production, we would like to stress the successive shifts that take place and 'where new layers are added while older ones always remain present. But once the research and innovation system comes under pressure, frictions between the layers become more apparent and some of the previous layers may gain prominence once more' (Felt et al 2013: 3). Recent developments, however, point to new challenges that raise questions about the relations between science and democracy. In times when scientific practices are more diverse, Ravetz (2022) argues that critical thinking about science itself is needed. If open innovation is driven by narrow commercial interests, whose interests will it serve and what new inequalities will it be part of shaping? What are the conditions and possibilities for objecting to representations that fail to be fair, just, or relevant, and who is responsible for responding to such objections?

The distance

When the relationship between science and politics is analysed by philosophers and social scientists – including those active in the fields of research policy studies and STS, who have this relationship as their specialty – the question of distance often comes up as a central issue (Jasanoff 1990, 2017). This also applies to all the actors whose practices are located in the borderland between science and politics. As already mentioned at the beginning of this chapter, researchers, politicians, and bureaucrats involved in science advice and science policy are focusing on keeping the two territories close to each other, but not too close.

However, Jasanoff notes a paradox in science advice in that distance is what gives the work legitimacy, while closeness generates a functioning working practice where things could be achieved in meeting places 'where scientific as well as political conflicts can be simultaneously negotiated' (Jasanoff 1990: 237). This possible paradox thus implies closeness in practice but distance in rhetoric (Hilgartner 2000).

Based on this reasoning, we can conclude that science advisers act strategically by performing a balancing act in managing the interaction between science and politics. In this vein, Thomas Gieryn (1983, 1999) suggests that science advisers carry out boundary work based on their professional interest and thereby strengthen their authority by maintaining both scientific integrity and policy relevance. The distance is negotiated and used strategically, based on who you talk to. Sometimes a large distance is advantageous; sometimes proximity is aimed for (Sundqvist et al 2015).

Jasanoff (1990), who has long studied environmental regulation in the US, distinguishes between research science and regulatory science (see also Demortain 2020). During the 1960s, a new era began in the industrialized Western world in terms of environmental regulation. The background was an increasingly intense debate on the environment and an engaged environmental movement that focused on environmentally hazardous industrial emissions linked to substances such as mercury, PCBs (polychlorinated biphenyls), and DDT (dichloro-diphenyl-trichloroethane), which were accumulating in food chains and causing extensive deaths among fish stocks and bird populations. Rachel Carson's book *Silent Spring* from 1962 became an important driver, as did the debate about the risks of nuclear power and radioactive emissions. According to Jasanoff (1990: 15–16), this debate centred on the conflict between technocratic and democratic risk regulation.

From a technocratic point of view, the knowledge needed and used in regulatory work must be linked to independent academic research that follows the norms of peer review. Whether a substance is environmentally hazardous or not should be determined by scientific knowledge. This means that technocrats make a clear distinction between science and politics. The great danger is politicized research. The democrats, for their part, believe that it must be made clear that environmental regulation is a value-based activity, which means that different voices should be listened to in regulatory work. Scientific knowledge is important but should not dominate.

According to Jasanoff, technocrats and democrats are partly right and partly wrong. Technocrats have a scientific view of knowledge, which means that scientific knowledge is independent and the most important factor in matters of risk regulation, and thus it is considered to guarantee the best result. However, Jasanoff's studies show that knowledge is not politically independent in regulatory processes. Like Nelkin (see Chapter 3), she believes that the political context affects the knowledge assessments made by experts as well as by those who have clear technocratic ambitions. On the other hand, it turns out that there is a strong majority in society who claim the independence of science, what Jasanoff (1990) calls the 'myth' of pure science. This myth is strong in American society, not least in the political administration concerning risk regulation (Jasanoff 1990: 236–237; cf Demortain 2020). Therefore, the democrats' demand for broader participation in regulatory work, to let more voices be heard, has weaker support. The public, politicians, and bureaucrats are too technocratic in their view of knowledge to allow this to happen. At the same time, Jasanoff finds in her studies that knowledge assessments are value based and that regulatory work is performed in close contact with political will and scientific assessments.

With the help of Jasanoff's work, we can thus distinguish between a front stage and a backstage in regulatory work, a difference in how the work is presented by involved actors and how it actually works (cf Hilgartner 2000). A technocratic risk regulation based on pure science is what is desired by many, and this is often how the work is also presented. In practice, however, scientific experts do exactly what they are not allowed to do according to the myth of pure science, namely addressing both scientific and political issues at the same time and trying to adapt them to each other (Jasanoff 1990: 237). If they did not do this, regulatory activities would not work. Unity and stability in risk regulation could not be achieved without close cooperation between scientific experts, bureaucrats, and political actors. This is expressed in practice in the meetings where regulations are formulated, even 'though their purpose is to address only technical issues' (Jasanoff 1990: 237).

Jasanoff (1990: 234) concludes that regulation is most of all about negotiation and strategic demarcation (boundary work), an activity that simultaneously assesses and negotiates scientific facts and political assessments, arriving at what is dangerous and not dangerous, what should be allowed and what should be banned, and implying the creation of the type of knowledge she calls 'regulatory science'. This knowledge is elastic enough to satisfy the research community's requirements for correct knowledge while at the same time being able to function practically to achieve stability and political legitimacy. To this is added the aim among involved actors of presenting regulatory work as more scientific than it is.

The direction

We have already noted in this chapter that the direction between science and politics, knowledge, and action is hardly simple or unambiguous. Many social scientists are also critical of the linear model of knowledge and argue that the interplay between science and politics always goes in both directions (Jasanoff and Wynne 1998; Beck 2011; Jasanoff 2017). Silke Beck (2011: 303) sums up this critique when saying that the foundation of the linear model is empirically incorrect: (i) more and better science does not always lead to increased certainty, and new uncertainties are often created instead; (ii) more and better science does not always resolve political conflicts, and instead more science can give more fuel for conflicts; and (iii) acting in accordance with the linear model paradoxically can lead to a depoliticization of politics and a politicization of science. This summary is in line with Nelkin's results on how experts in the political arena often increase political conflicts when trying to solve them with the help of expert knowledge (see Chapter 3).

The democratic approach that Jasanoff presents means that more voices than just those of experts should be heard, and this is particularly relevant when it comes to risk and environmental regulation. This applies not least to groups that have been affected or will be affected by the substances or activities that are to be regulated. When it comes to extended participation in this type of process, metaphors that indicate flows and direction have been common as a complement to those that indicate distance. The public can, for example, be involved downstream or upstream (Wilsdon and Willis 2004), and participation can be thought of as a ladder where citizens gain more influence the higher up on this they move (Arnstein 1969). Upstream involvement means that the public is involved at a stage when neither policies nor decision alternatives have been formulated, while downstream involvement means that participation takes place in a phase where many decisions are already taken (Stilgoe 2007: 33). One problem, and in line with the deficit model, is that experts seldom want to share their power with a wider public, which is often considered ignorant. Attempts to involve the public in decision-making processes are in contrast to the deficit model presented as upstream engagement and as taking place at the top of the ladder, but in reality, they are often about experts providing information to the public, rather than the latter being given opportunities to influence (Tait 2009). The questions of how meetings between experts and the general public are organized and framed becomes central and is about what quality we can find in participatory processes.

From the early 2000s, we can see an emerging democratization in terms of various decision-making processes (Fishkin 2011). The need to strengthen public participation was explicitly stated by the European Commission in 2001 in its 'principles of good governance', which emphasized, among other things, that political decision-making must be more open and inclusive (Commission of the European Communities 2001). At the same time, the role of experts in informing the public is emphasized together with the importance of restoring public confidence in expert opinion and making expertise more open and accessible.

'Recent food crises have highlighted the importance of informing people and policy makers about what is known and where uncertainty persists. But they have also undermined public confidence in expert-based policy-making' (Commission of the European Communities 2001: 18). This quotation from the European Commission's White Paper on European governance highlights that the relationship between experts and the general public is central and, at this time, also in a critical state.

Since the end of the 1990s, public transparency and opportunities for participation in the field of environmental policy have been strengthened through international conventions and EU directives. The Aarhus Convention of 1998 established citizens' right to information, influence, and justice in the environmental field. Other examples of regulations that in various ways strengthen the public's right to information and participation are the EU directives that regulate hazardous substances (Seveso Directive from 1997) and water management (the Water Framework Directive from 2000).

When citizens protest or otherwise express distrust of experts' assessments regarding environmental impact, health issues, technological development, or other important societal issues, the overlap between science and politics becomes apparent. Those who exercise expertise are dependent on public trust, and when mistrust is expressed, there is a need to bring the public closer to expertise and expertise closer to the public. These activities are about both distance and direction.

The language of open innovation and science is somewhat similar to the language of participatory governance, as the aim is to 'engage and involve citizens, civil society organizations and end users in co-design and co-creation processes and promote responsible research and innovation' (European Commission 2019). The regulation that aims to strengthen public transparency and participation relies on improving concerned publics' right to information and consultation. It assumes a difference and distance between the developers/decision makers and the concerned publics. The language of open innovation seems to suggest proximity – a close relation between science and citizens. However, innovation is also connected to strong ideas in a certain direction – that is, that innovation needs to be scaled up – and as such it is also 'linked to the creation of widely shared collective visions about the future, which makes these visions a key battleground' (Pfotenhauer et al 2022: 9).

An important question then becomes in what ways these visions are opened up for a wider public, or whether citizens are only invited to take part in processes that are pre-framed and, thereby, are given instrumental roles to pursue already defined goals. Concrete processes of open innovation that allow for citizen involvement, then, 'not only [frame] collective issues and [define] how to act on them, but also [distribute] the ability to intervene on these issues in unequal ways' (Pfotenhauer et al 2022: 20). Thus, the question of direction suddenly becomes irrelevant when issues and forms for participation are intrinsically interwoven and constitute each other. This further points to the need to critically reflect on the qualities of scientific practices when these become more and more diverse: How trustworthy are they? What purposes or goals motivate them? What issues and concerns are included and excluded?

Boundary work and boundary organizations

The overlaps between science and politics come in many variations, as STS scholars show with the notion of *boundary work* (Gieryn 1995, 1999). The boundary, or boundaries, between science and politics are not given. Work is required to maintain the boundaries – for instance, in relation to distance and direction – but also to bring about interaction and exchange between different institutions, organizations, types of knowledge, and scientific disciplines. This means strategic work that, among other things, aims to achieve epistemic authority in political decision-making.

To be an expert is to be an expert in an area. Based on this idea, the question of the area's domain and jurisdiction becomes central. The purpose of boundary work is to strengthen authority by making clear who has the competence to express what. Boundary work is thus about actors, issues, and subject areas. When boundaries are drawn around a certain expert domain, it is made clear what issues the expertise deals with (and what issues it does not address) and who can and should participate in this work. But the work applies not only to the boundaries between different areas of knowledge but also in relation to what issues are not seen as 'knowledge issues', but rather understood as 'political' and 'democratic issues'.

The concept of boundary work provides opportunities to study the limits of science from the perspective of actor-based practices. The boundaries should therefore be seen as changing and as the subject of negotiations and conflicts where involved actors act strategically. 'Science' and 'politics' become labels used by actors to describe and protect what they do, in a way that suits their interests. Science and politics, and not least the boundaries between them, become conditional in relation to the actors' practices and social contexts.

In many political areas, knowledge is highly valued. Regarding environmental issues, knowledge is considered something indispensable for drawing attention to problems, but also for politically regulating environmental risk or the use of natural resources. What amounts of environmentally harmful emissions should be tolerated? How and when should scientific knowledge be used in the political decision-making process? In this regulatory work, we find strong arguments for both separating and integrating science and politics, and not seldom a desire to achieve both at the same time (Sundqvist et al 2015). Organizations that act in the scientific as well as the political field have been called 'boundary organizations' (Guston 1999). These mobilize scientific knowledge for political purposes and are characterized by the fact that they seek an interplay between science and politics within one and the same organization. These boundary organizations, or hybrid institutions (Callon et al 2009; see also Beck 2012), handle knowledge issues from a political point of view (Corner and Groves 2014: 744).

Government authorities could exemplify boundary organizations, but there are also a few research institutes and private organizations which act as lobby organizations and place knowledge at the forefront of political activism (Garsten et al 2015). The IPCC is an interesting and important example in this context. The panel, established in 1988, is an intergovernmental UN organization and has the task of providing the world's decision makers with a solid scientific knowledge base for what is happening to the world's climate (Agrawala 1998; Bolin 2007). The IPCC summaries of the research situation in the climate field (assessment reports) are published regularly in cycles of six or seven years. The first report came out in 1990 and the sixth and most recent in 2021–2023 (comprising three different sub-reports from three working groups and one synthesis report). The IPCC can be understood as the scientific part of a collaboration in which the United Nations Framework Convention on Climate Change (UNFCCC), signed in 1992, is the political counterpart. The UNFCCC is the international negotiating organization for climate issues, which has resulted in the Kyoto Protocol of 1997 and the Paris Agreement of 2015. The IPCC reports form an important knowledge base for these negotiations. Through these two organizations, a division of labour between knowledge compilations and political negotiations is created at the global level (Miller 2004).

The IPCC aims to present knowledge that is 'policy relevant, but not policy prescriptive' (quoted in Yamin and Depledge 2004: 466). This wording, which is often emphasized by the organization, means being close to politics but not too close. The IPCC wants to be seen as a scientific organization which summarizes scientific research, but it is important to note that its raison d'être lies in the fact that the summaries will lead to policy measures to reduce greenhouse gas emissions and negative effects of a changing climate. Furthermore, as already mentioned, the IPCC is politically controlled. Government representatives from UN member states together with researchers nominated by their governments decide on the outline of the reports and evaluate draft reports (De Pryck and Hulme 2022). It has been said that the work of the IPCC 'is unique in the intergovernmental arena in its combination of expert scientific analysis with government review and negotiation' (Yamin and Depledge 2004: 475).

The IPCC is an important example of a deliberately organized collaboration between government representatives and the global research community based on the aim of synthesizing policy-relevant knowledge, and, as such, it is a boundary organization. But this international work is far removed from ordinary citizens, who are not included at all. In the field of the environment, there is a variety of examples of the interplay between science and politics, with citizens sometimes included and sometimes not included in these activities. With the help of metaphors involving proximity, distance, direction, and ladder, and with concepts such as boundary work and boundary organizations, we can carry out empirical studies of what forms these interactions take and, not least, what consequences different boundaries and forms of inclusion have for different groups.

Hybridization

When things cannot easily be classified in *either/or* terms in the divisions between science and politics, or between nature and culture, they are hybrids. Anthropogenic climate change is a hybrid phenomenon, as are frozen embryos, nuclear waste, and plastic pollution. Paradoxically, modernity's efforts to separate nature from culture, and science from politics, is aligned with extensive creation of hybrids. The ecological crisis is a matter of hybrids out of control. It is due to the belief that nature is something external to us humans in society that practices such as splitting atoms, gene modification, mass production, and mass consumption have been established. According to Latour (1993), modernity is all about the separation between nature and society, and therefore ecological threats have become the defining element of our contemporary world. This happens when negative 'side effects' appear and it becomes clear that these could not be separated and regulated, and so potential dangers are no longer kept at bay. When we are dealing with complex issues such as climate change, practices of separation will become more difficult to uphold since there are uncertainties, issues concerning ethics, and injustices that escape narrow expert framings. In an era of ecological crises, the distinction between spokespersons of humans and spokespersons of non-humans is becoming increasingly untenable, Latour (2004) argues, presenting his vision of democracy as based on the creation of a 'parliament of things' in which humans and non-humans may come to be represented together.

To admit hybrids also means to admit conflicts. The ecological crisis cannot be solved by pedagogical means – that is, by communicating correct science in a better way to the people who can act accordingly. Rather, the ecological crises must be treated as what they are: political struggles (Latour and Schultz 2022). Understanding hybrids and hybridization in this way means accepting the ecological crisis as inevitably bound up with conflict. Therefore, it is reasonable to ask: What organizations dealing with science–policy issues in our societies can afford to fully appreciate this understanding of hybridization? What time is set aside for acknowledging all parties in a conflict, and how will conflicting perspectives be presented and discussed? Can we expect that established institutions and organizations will initiate a public exposure of conflicting views, or is it more reasonable to expect that they will defend a separatist view?

Conclusions

In this chapter, we have presented approaches that are based on an understanding of science and politics as separate practices and institutions but have a focus on how to improve the interplay between science and politics. The discussions of science policy and science advice have a strong practical relevance, since they are about how science is organized in society and also how scientific knowledge is taken advantage of. These discussions are part of an ongoing academic discussion on how to best understand, study, and explain the interplay between science and politics, and sometimes practical and theoretical ambitions meet, as, for instance, in the work of the European Commission's SAPEA.

The examples given in this chapter complicate the picture of a clear division between science and politics. Rather than suggesting a clear-cut separation, we find efforts to balance, bridge, and facilitate specific relationships characterized by proximity or distance, and as a result, specific mixes between science and policy are accomplished. The usefulness of science is not a simple matter of demand by decision makers; rather, the interaction goes both ways. Active work from both sides shapes how particular supply and demand relations look – for example, scientists may package the results in a particular way and present their knowledge as something useful. Science advisers, by definition, have left their laboratories and started to interact with other actors, such as decision makers, who, for instance, by funding research, also shape the supply side of science. How these processes are organized also refers to the actors involved; new formats for citizen involvement, for example, cannot be seen as external to the knowledge-producing processes they are involved in, but rather are intrinsically related to and constituted in relation to the knowledge production in question.

In the approaches of Mode 2, post-normal science, and boundary organizations, we meet similar ambitions about studying today's situation of complex problems in need of scientific understanding as well as political decisions. While the first two are clearly normative, trying to show how to manage this mixed situation of scientific and political challenges, the notion of boundary organization is more descriptive as it analyses how boundaries are drawn and how differences as well as combinations are constructed in attempts to establish connections between science and politics. Current developments in how knowledge is produced can be understood as an escalation of Mode 2, drawing on digitalization and new efforts to facilitate open innovation and science. However, simultaneously, as there is constant hybridization, we find strong defenders of separatist views. Why is this? It is because accepting hybridization also means the acceptance of conflicts and political struggles – something that many actors want to avoid for both practical as well as political reasons. To uphold struggles over knowledge production – that is, to publicly interrogate its goals, perspectives, and societal implications – is simply more in the interest of those actors and interests that are excluded, non-recognized, and not yet represented.

PART III

Co-production

5

Co-production of Scientific Knowledge and Societal Order

Introduction

This chapter goes deeper into the complex relation between science and democracy. Previous chapters discussed how science and politics are separate and that one elite (scientific experts) is delegated the power to represent nature (as knowledge objects) and another elite (decision makers, not only elected politicians but broadly speaking) is delegated the power to represent the people. This double delegation (delegative democracy) is based on a distinction between what is represented and who can represent. These two forms of representation divide the world into the two domains of knowledge and politics and, in addition, create a sharp division between those who are in power (the two elites) and those who are not (the ignorant mass). However, these separate domains are also interlinked, and as we have seen earlier, the boundaries between them and their authority and legitimacy can always be questioned and change over time. The separation between science and democracy, between representatives and represented, is a joint product. One of the most important ideas within STS research is that knowledge and social order are intertwined or, as it is often expressed, *co-produced* (Jasanoff 2004b).

Science and democracy are not independent from each other. Democracies legitimize and back up decisions and reforms with expert knowledge, and an uneven distribution of knowledge and education in a society is seen as a democracy problem (Sismondo 2010: 80). Governing requires knowledge, and science is intermingled with power. Science and democracy may be the result of a joint process, but they are often presented as separate, which has consequences for how practices are performed and leads to what they have in common being concealed. Knowledge is understood and presented as being independent of social order and power, especially in the scientific community. Researchers have a professional interest in keeping sharp distinctions between scientific knowledge and other types of knowledge as well as sharp distinctions between scientific knowledge and political interests (Gieryn 1983). At the same time, researchers also have an interest in getting attention from outsiders and they want their results to be used outside of the research community. This duality, which we have discussed previously, implies the wish of researchers to have, simultaneously, distance and closeness to politics. What we present and develop in this chapter is the idea of co-production, which means that the relationship between science and politics is not only a question of distance, or a balancing act between distance and closeness, but a more entangled relation that means that science and democracy presuppose each other. They are co-produced.

The reason why science and democracy are often presented as separate is that modern societies are based on the idea of differentiation and autonomous institutions (see Chapter 2). The best representation of nature (knowledge objects) can be achieved when science works as an autonomous institution. But what does this autonomy mean? Science as an institution is autonomous if it is free from direct political interference as well as from other types of societal influence, such as values, religious beliefs, political opinions, and the researcher's own life conditions and personality. Autonomous science is not influenced by extra-scientific circumstances. STS research has been devoted to showing that this separation between the inside and outside of science is only half the truth. Latour (1993) has expressed this by pointing to the two parallel processes that characterize modern society: *purification* and *hybridization*.

In this chapter, we first develop the idea of co-production, and thereafter we follow up on Latour's conceptual pair of purification and hybridization, which has close affinity to the notion of co-production but offers the possibility of studying separation and co-production as processes that in practice are developed simultaneously, or better: separation takes place within the frame of co-production and is thereby a result of processes of co-production, as are other kinds of combinations of science and democracy. In connection to this, we also discuss Callon's ideas about how scientific controversies that are also political controversies – that is, mixed controversies – can be managed through cultivated 'hybrid forums'. The aim of Callon's approach is to avoid the defence of a strong separation between science and politics as well as between experts and laypeople by acknowledging the hybrid character of issues. This is an alternative to understanding modern society as delegative democracy (see Chapter 2). After this follows a section on the role of experts in acknowledging hybridity, and we argue that what is required is that experts attend better to the limitations of their own competences. The idea of reflexive expertise then becomes a complement to the deficit model (see Chapter 3), one which identifies deficits not only among groups of the public but also among experts. Finally, we discuss a positive versus a negative view of politicization of science.

Co-production

STS researchers often adopt an agnostic attitude to the question of the boundaries of science. The idea that the boundaries of both science and politics – their cultures and practices – are in principle open and changeable implies that these boundaries can be empirically studied (Gieryn 1983, 1999). Yet, for STS analysis, it is not only how boundaries are established, defended, and negotiated that is of interest, but also how the different sides of the boundaries are co-produced (Latour 1993; Jasanoff and Wynne 1998; Jasanoff 2004b; Hilgartner et al 2015).

Co-production of scientific knowledge and societal order means that scientific knowledge both shapes and is shaped by and 'embedded in social practices, identities, norms, conventions, discourses, instruments and institutions' (Jasanoff 2004c: 3). The idea of a co-production of scientific knowledge and social circumstances is as old as research in STS. For good reason, it could be argued that STS implies co-productionist thinking. Early in his career, Harry Collins (1992) argued that the development of expert knowledge takes place together with group formation. Behind scientific agreements, we find group mobilization, and these two things are mutually dependent (Callon and Law 1982). However, what often happens is that the social aspects are ignored or forgotten when knowledge has been established as agreed and true knowledge.

Jasanoff (2004a: 18f) distinguishes between two main strands in STS research that focus on related yet different aspects of co-production. She calls these strands 'constitutive' and 'interactionist' co-production. The constitutive strand is about the fundamental societal processes that create the reality we live in; here, when something is categorized as 'nature', demarcations are also made as to what belongs to 'society'. The fact that the chemical elements and the human genome are understood as nature, means that they are separate from what is understood as society, politics, and culture. Research belonging to this strand includes Latour and Callon's actor network theory (ANT) and their understanding of the overarching processes of purification and hybridization (see Latour 1993). Another approach that belongs to this strand is Donna Haraway's understanding of how humans and non-humans become what they are in relation to each other. In her relational ontology, Haraway (2008) uses the notion of 'becoming with'.

The other main strand, which Jasanoff calls 'interactionist co-production' (Jasanoff 2004a: 28), is research that analyses scientific and mixed controversies against the background of already established societal orders. Belonging to this strand are, for example, symmetrical analyses of scientific controversies and conflicts between experts and non-experts. Here we find studies that focus on 'knowledge conflicts within worlds that have already been demarcated, for practical purposes, into the natural and the social' (Jasanoff 2004a: 19).

The emergence of modern science and democratic rule as important institutions in Western modern societies can be thought of as the first order of co-production as discussed here – that is, *constitutive co-production*. Science and democracy have common roots in the Enlightenment. The development of democracy (specified versions of democracy) has thus occurred in alliance with the Scientific Revolution, which in our words means that the authority of scientific knowledge and of democratic governance are legitimated by *representation* (see Chapter 1).

Within the societal order that characterizes modern society – and in which science and democracy are important institutions – the relationships between science and politics can differ. This societal order, understood as constitutive co-production, implies that fundamental issues concerning what 'is' (how nature and technological artefacts work and what risks and challenges they bring with them) and how they 'should' be (how society should support and use science and technology) are intrinsically intertwined (Jasanoff 2012: 8, 16).

The knowledge that is put forward as relevant for solving a particular problem will affect and reshape (co-produce) the problem formulation. This intricate weaving of problems and solutions can be made visible when looking closer at regulative work. Rather than seeing science as a direct and external input into decision-making, we need to understand how science and expert knowledge are being adjusted (translated) to the context for which they are to be used. This is an interplay in which knowledge is modified and the problems that are to be solved are made concrete. As the solution and the problem are formulated in tandem and in relation to knowledge, it is impossible to say that knowledge precedes decisions, or that science one-sidedly affects politics. Such processes of co-production point to the faulty and misleading ideas that underpin the linear model of knowledge (see Chapter 3).

We can see an example of interactionist co-production in how governing elites responded to the protests against the establishment of 3G in Europe at the end of the 1990s and beginning of the 2000s (Soneryd 2007). In Sweden, a country in which protests were prominent, citizens protesting against the planned infrastructure for 3G mobile telephony had a range of motives for their resistance: aesthetics, threatened local governance, environmental and health impacts, and unclear costs and benefits for different groups. Municipalities and local governments reacted in different ways to these protests. While the protests were multifaceted, the responses to them were rather narrow. The Swedish Radiation Protection Authority (Strålsäkerhetsmyndigheten – SSM), on the basis of its mission to regulate risks related to radiation and electromagnetic fields, was the government authority that primarily responded to the public reactions. In this case, the problem formulation that was the official and dominant one was tightly bound to a specific knowledge production, and engaged citizens could be dismissed as a group with a deviant (and wrong) risk perception: they did not understand radiation issues correctly. The problems and solutions were thus co-produced with specialized expert knowledge.

Nuclear waste is another example. Nuclear waste can be seen as a case in point when it comes to hybrid issues – that is mixes between technology, nature, and society. What nuclear waste is, and how it can be managed in a safe way, is often assessed as being a technical issue, while questions concerning public acceptance of or resistance to a repository for nuclear waste are seen as social and political issues (Sundqvist and Elam 2009). But these aspects are inseparable and affect each other. Nuclear waste disposal is a hybrid issue and a result of co-production between politics, technology, and geology. During the 1980s, the Swedish Nuclear Fuel and Waste Management Company (SKB) was met by local resistance and intense protests against the test drillings the company were conducting and planned to conduct when searching for the best bedrock for the waste to be stored. SKB's process for finding the safest method and best place for a final repository was at the time very much connected to the question of finding the right geological conditions.

The project failed since the company and its experts did not include the local population. The intense protests showed that the alliance between experts, the bedrock, and the drilling equipment was not enough to create a safe geological disposal of nuclear waste. The ambition to create a stable relation with the bedrock neglected the necessary involvement of local decision makers and citizens, who had to be convinced about the location of a final repository in their region. Hence, SKB changed its strategy. The focus on the bedrock was played down while societal factors, such as local involvement, were given more attention. To achieve greater flexibility in relation to the siting of the waste, the bedrock was reduced to one of several barriers that, together with encapsulation in copper canisters surrounded by bentonite clay in the deep bedrock, would keep radioactive waste separate from society.

The example shows how resistance led to a changed expert assessment about what knowledge and technologies were needed to take care of nuclear waste in a safe way. Through this new focus, a novel mix of components was presented and a hybrid technology was created, orientated towards finding a combination between a bedrock that was 'good enough' and a municipality that was 'positive enough' to accept the nuclear waste within its borders (Sundqvist and Elam 2009: 17). Despite this hybridizing strategy, that aimed to create alliances between nature (the bedrock), technology, and the population, SKB (the experts) continued with processes of separation: on the one hand, when they present their own activities, they talk solely about these as technical; on the other hand, they expect others (a concerned public) to stick to 'social issues' and to not question safety (because in a separated world, safety is all about nature and technology). SKB considers hybridity as something the company should deal with while local politicians and citizens should focus on social issues, such as social and economic benefits. For the company, this means transforming local attitudes from rejection to acceptance. Hybridity strategies are kept backstage in the work performed by SKB, while separation is what the company presents in public (cf Jasanoff's discussion of the myth of pure science in Chapter 4).

The idea of constitutive co-production does not exclude the possibility of separating science and politics (see Figure 5.1). As we have already seen, many actors in modern society expect a clear separation. For example, the demand for policy-relevant science is often connected to ideas about a 'pure' science according to Mertonian ideals of disinterested science (Jasanoff 2012: 162). Separation can thus be seen as an effect of certain actors' presentations of particular phenomena as given 'facts' (and, hence, free from social interests), since they have a strategic interest in doing exactly that. Figure 5.1 illustrates constitutive co-production, which we presented already in Chapter 2, in relation to the concept of delegative democracy that relies on the distinction between two elites and an undifferentiated ignorant mass. With Jasanoff's distinction between constitutive and interactionist co-production, however, we can also see variations in what interactions between the elite groups and concerned publics and citizens look like. These interactions differ depending on the particular policy area and its history, regulatory contexts, and specific problematizations, which vary over time. The figure illustrates that this interactionist co-production involves practices of separation as well as hybridization.

A co-production approach does not explain knowledge with the social, in the way social constructivists explain scientific knowledge through social interests (and actor strategies). The social constructivist approach suggests an inverted linear model and is, hence, based on a clear division between the cognitive and the social, science and politics. In addition, the focus is on how science can be explained through social factors (see

Figure 5.1: Constitutive and interactionist co-production

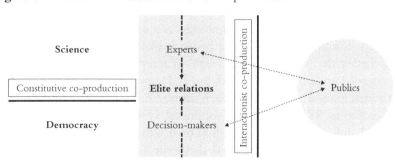

Chapter 3). Co-production, on the contrary, suggests that science and politics simultaneously shape each other in a joint process.

In other words, while there is a diversity of ways in which science and politics can relate to each other in practice, the ways they relate are always a result of co-production. Their relations are not static, but changeable and subject to strategic action from actors who gain from either changing or stabilizing existing forms. Studies of co-production of science and politics need to be undertaken with openness to the fact that there are many forms of expression. Established forms can be critically evaluated from the assumption that alternatives are possible (Lidskog and Sundqvist 2011: 349–350; Jasanoff 2017: 261).

Hybrid forums and technical democracy

Problems and conflicts often appear because society is organized and understood as separate even though it consists of hybrids. If we take hybridity seriously, this means that the capacity to act (agency) can never be understood solely through single human subjects or individuals. Agency is always the result of assemblages or collectives of human actors and things, groups and infrastructures, that are mobilized and shape new forms of alliance to accomplish action. New assemblages also create new identities, objects, and subjects. The compositions and capacities for action that are generated through these are important themes for ANT research. To study these assemblages means studying processes of co-production.

When scientific knowledge is disseminated throughout society and becomes relevant to many actors, *overflows*, to borrow a term from Callon (1998), frequently appear. Scientific facts are mixed with concerns expressed by the public. Actors other than researchers interpret scientific facts and find them useful, but they also find that these facts will not give complete answers, or relevant answers, or that the answers given hide dimensions that are worth acknowledging. Overflows happen when things which have a clear meaning within the walls of the scientific laboratory are placed in new and more complex contexts. Overflows should thus be understood in relation to the concept of *framing*. Framing means that some aspects of a particular issue are seen as relevant, while others are deemed irrelevant and ignored. Framing is a way to simplify, categorize, and order the world. If the framing is accepted, there are no overflows. Overflows are signs that the framing is questioned (Callon 1998: 17). Economists use the term *externalities* for this phenomenon, and sociologists term it *non-intended consequences*. Overflows generate events or dimensions that were not intended or foreseen beforehand, but are acknowledged at a later stage. When this happens, the framing is revealed as a simplification and can then be criticized for being inadequate and insufficient when positioned in a new context and assessed by new actors.

Overflows mean that the situation becomes 'hot', the consensus around the framing ceases to exist, controversies arise, and a multitude of actors have opinions about how the issue should be framed or reframed (Callon 1998: 260). This is the situation when the ownership of problem formulation and suggested solutions becomes blurred, in contrast to the situation in the laboratory, where experts are in control. Established divisions between science and politics, and experts and publics, simply do not work anymore. 'Overflows are inseparably technical and social, and they give rise to unexpected problems by giving prominence to unforeseen effects' (Callon et al 2009: 28).

According to Callon and his co-authors, this means that existing scientific and political institutions that are used to manage 'pure' and separate issues are unprepared for managing these new mixed objects, which are at the same time scientific and political. The scientists in the laboratory might understand overflows as resistance to their projects and feel threatened, and thus try to maintain their assumptions and original framing. However, Callon and colleagues argue that this just increases conflict and therefore the responsible institutions must be changed. The established political institutions must 'be enriched, expanded, extended, and improved' to make 'our democracies more able to absorb the debates and controversies aroused by science and technology' (Callon et al 2009: 9).

A consequence of this is that we need new ways of organizing decision-making processes for controversial issues around science and technology. This can be accomplished only by crossing the deeply entrenched divisions in Western modern societies, between laypeople and experts, and between citizens and their representatives (Callon et al 2009: 35). Delegative democracy is questioned. To defend strong divisions in overflow situations will only heighten the conflicts. Overflows in relation to scientific knowledge, when it is diffused in society and has consequences for people and things, cannot be addressed adequately by maintaining the old divisions. The only way to solve the conflicts is to take them seriously and to understand them as necessary for managing within a new type of institution, which Callon et al (2009) call the 'hybrid forum'. Situations of conflict must be acknowledged, and the limits of expert knowledge as well as the relevance of laypeople's knowledge must be acknowledged. In addition, uncertainties and conflicts need to be discussed openly in situations of extended participation that include experts and publics alike. This situation should be understood as hybrid, in which hybrid actors manage hybrid objects.

Callon expects that societal controversies around science and technology are many and serious, and emerge continuously. He summarizes this by arguing that 'not only are "hot" situations becoming more commonplace [but] it is becoming exceedingly difficult to cool them down' (Callon 1998: 262; see also Sundqvist 2014). For Callon, it is important to get to know this new situation, which cannot be solved by more certain science according to the traditional mode. To see the new situation as a possibility and not a threat is a first important step towards a solution. It is only when responsible authorities, or other actors with power to change the situation, acknowledge issues as hybrid that controversies and overflows can be managed and robust solutions can be presented.

> Controversy allows the design and testing of projects and solutions that integrate a plurality of points of view, demands, and expectations. [...] [I]t allows laypersons to enter into the scientific and technical content in order to propose solutions, and it leads the promoters to redefine their projects and to explore new lines of research able to integrate demands they had never considered. (Callon et al 2009: 32, 33)

It is, however, important to note that Callon does find it appropriate to delegate some issues to technical experts. In a 'cold' situation – that is, in a situation with no open conflicts or known overflows – it is enough to involve experts (Callon 1998: 262). To understand a situation as 'hot' – as controversial and characterized by a mix of expert knowledges and lay knowledges – does not mean that solutions cannot be sought. But solutions cannot be found in scientific certainty. It is impossible to give room for the unrepresented once and for all, since new overflows might appear. Instead, the solutions are about searching for 'preliminary negotiations' (Callon 1998: 263) and 'collective experimentation' (Felt and Wynne 2007: 67, 71). This means that the first and important task for both analysts and practitioners is to make judgements about when a situation is 'cold' or 'hot' (Sundqvist 2014). According to Callon, this is most easily done through searching for overflows.

We noted already in Chapter 2 that Callon sees nuclear waste as an important example of how sharp boundaries between science and politics were established. Delegation to experts and the establishment of two elites have been clear patterns in the issue of nuclear waste, and hence the shaping of an ignorant public. This ignorant public is, according to Callon, 'no less a by-product of nuclear power than is radioactive waste' (2009: xiii).

Social science has supported this division and manifested the existence of an ignorant public. It is only after studies are carried out of the public and its knowledge, or knowledge deficits, that the 'ignorant public' starts to exist. Through surveys and opinion polls, it is possible to show that 'this "public" is increasingly ignorant and less interested in decisions' (Callon 2009: xxi). In this way, social science has supported the existence of Plato's cave dwellers (see Chapter 2). In debates over nuclear power since the 1970s, the ignorant public is constantly referred to. Callon and his colleagues argue that nuclear power is not only a case in point when it comes to a separated world, but also the main example of a hot situation. It is because of the sharp divisions that the issue is particularly vulnerable to overflows. The harder you push for a particular framing, the more demanding it is to make it hold, and the greater the risks of overflow. The waste generated from the nuclear reactors is a clear example.

> We thought that geology would ensure a decent and definitive burial for nuclear waste that everyone would respect, and now wine growers, whose voice had not been heard, are worried, not about the effects of radioactivity, but about far more worrying commercial effects, since they are in danger of losing foreign customers who could take fright on learning that the grapes ripen some hundreds of meters above containers filled with nuclear substances. [...] Questions that were thought to have been settled definitively are reopened. Arguments multiply and the project constantly overflows the smooth framework outlined by its proponents. (Callon et al 2009: 9, 15)

What the solution to the nuclear waste issue will look like depends on whether or not old divisions are defended. We do not have a particularly good reason to believe they will be abandoned. There are many experts, industry actors, and decision makers that have an interest in keeping the boundaries and maintaining the separation between experts and laypeople. This means that there is a constant risk of overflows arising and unproductive conflicts continuing (Barthe et al 2020, 2022).

Callon's argument implies that an expanding room for hybrid forums should be created in the gaps that have appeared in the classical divisions between laypeople and experts and between elected and electorate. These divisions have shaped our current societal institutions. This means that an established space for hybrid forums is often missing and that there is a lack of institutional support for issues that need to be discussed as technical and social at the same time. Although the issues are 'hot', it is much easier to act as if they were 'cold' and to continue to separate technical spheres from political and to categorize the actor groups that belong to each sphere (experts, politicians, laypeople). Spontaneously emerging hybrid forums due to overflows, which we can see in public media discussions or among mobilized citizen groups, are difficult to handle through existing institutions, even though this would lead to more robust decision-making. What often happens is that efforts are made to cool down the hot issues, which means that the old traditional divisions are maintained (Sundqvist 2014). According to Callon, this is not a solution, but rather something that tends to heighten the conflicts.

Noortje Marres (2007) argues that overflows imply the acknowledgement of issues that so far have been unrepresented. Overflows mean that something

new emerges and defies the established order (Callon 1998: 17). In order to gain any influence, these new unrepresented issues must be understood as a problem also from the perspective of established institutions. Since these institutions will typically not discover or acknowledge such issues themselves, they are often made visible by concerned publics who emerge around the issues. The unrepresented issues may concern uncertainties, ethical aspects, injustice, industrial pollution with environmental and health impacts – anything that previously has been ignored by experts and other elites. In this way, concerned groups function as correctives to the mistakes and blind spots of responsible actors, such as industries and government authorities. The reactions of governing bodies to these kinds of protest are crucial for our understanding of how democracies work. Societal institutions can react by including the new issues, ignoring them, or denying that the unrepresented issue is important enough to include. The opinion groups and protests that arise in relation to various societal issues should, according to Marres, be understood as overflows, and those who articulate problems that need to be managed should be viewed as articulating issues that need to be represented. These groups are therefore a positive democratic force that can have the function of extending democracy and repairing it when it does not work.

What influence these concerned groups of the public have on how issues are managed is an open question, and it is related to the question of who is seen as a legitimate participant or representative for particular issues or problems. The mere existence of unrepresented phenomena, however, provides an opportunity for a public who is silenced by double delegation (as ignorant cave dwellers) to mobilize and actively take part in transformative work (cf Chapter 2). This is important, since an engaged public often arises as a reaction to the fact that people have been affected in a negative way by the consequences of governing bodies' decisions. In this way, representations of issues are produced in processes that simultaneously shape the actors and their identities.

Reflexive experts

To protect ourselves from the nightmare of populism, Collins and Evans turn against a too far diffusion of epistemic issues – that is, they seek to level out epistemic differences (see Chapter 3). As Collins (2014) puts it, all of us are *not* scientific experts. Collins and Evans believe that in a society that takes both science and democracy seriously, we must allow ourselves to separate them and recognize that there *are* issues where specialists have better answers than the general public. But there are also issues where experts lack knowledge, where specialist knowledge is sometimes of no value, and where experts can cause damage if they are allowed to influence alone in a narrow-minded way. There is, thus, a need for continuous reconstruction in which we are constantly observant and examine what rightful place the scientific expertise should have in relation to a specific issue. This is a difficult task, as the relevance of expert knowledge varies and individual questions change over time: knowledge changes, new facts can be added, and the same knowledge can be assessed differently in relation to legitimacy and relevance and for various groups.

Wynne (1987, 1993, 2001) has, for a long time, studied the authority of scientific expertise. He agrees with Collins and Evans that the role of expert knowledge in a democratic society is a key concern, but he opposes the view that this is only a question of knowledge. Inspired by Nelkin's studies of experts (see Chapter 3), Wynne points out that laypersons could have knowledge that is of crucial importance in order to judge the role of expertise. For Collins and Evans, consensus among experts is the important criterion, including in political contexts. For Wynne, however, this is *not* the most important criterion, and he argues that laypeople are often more apt to judge the relevance of scientific knowledge than experts are. Experts often have a strong interest in giving much, and even too much, weight to expert knowledge and, hence, a tendency to exaggerate the role of expertise.

Similar to Callon, Wynne puts much emphasis on how a question is framed. Scientific experts have a strong tendency to reduce the issue in question, to strip it from everything irrelevant and in this way make it controllable, calculable, and possible to quantify. The public is more interested in extending the issue and focusing on what makes it important, where its risks lie, which groups may be affected, and who may benefit – for example, in relation to new-build nuclear reactors (see also Jasanoff 2003: 240). These understandings of issues are difficult to calculate and, therefore, they go beyond what the technical experts are specialized in. But questions of this type are marginalized when it is *scientific consensus* that is the main criterion for giving experts a crucial role in decisions about controversial science and technology.

In the following, we describe how Wynne uses a co-productionist approach when studying issues he considers as having a mixed (or hybrid) character by simultaneously focusing on how issues are framed in a mutual process where actors and actor identities are also produced.

With the help of the deficit model (see Chapter 3), Wynne focuses on what Callon calls 'the undifferentiated, sluggish, ignorant public' (2009: xii; cf Plato's and Latour's discussions of the cave dwellers in Chapter 2). According to the deficit model, ignorance is relational. Ignorance arises in relation to someone who is knowledgeable, as a lack compared to someone who knows. Wynne argues, similar to Callon, that such shortcomings arise when a division is made between experts and laypeople, which puts experts in the position of being the norm from which to judge others. Laypeople are then defined based on their lack of scientific literacy and (mis)understanding of how science works (Wynne 1995; Irwin and Wynne 1996).

Scientific knowledge is always a reduction and simplification of a complex reality. Simplifications are the strength as well as the weakness of scientific knowledge. Research at the time when nuclear power was first being developed focused on splitting the uranium atom – that is, nuclear fission – and doing this in a controlled way and on a large scale for heating water to power turbines and generate electricity. The focus of this research was thus never to come up with clear answers to questions concerning the risks of nuclear fission or what to do with the by-products of this activity, and it was certainly not about how many nuclear reactors – if any – that a specific nation state would need. The researchers were occupied with uncertainties and risks around the issue of splitting the atom and being able to control the chain reaction that occurs when this releases neutrons that cause more splitting and generates more heat. The speed of this chain reaction needed to be controlled with neutron-absorbing elements to avoid nuclear explosion. These were, and still are, technical challenges.

When nuclear power is discussed in public among a wide range of actors and with a much broader agenda, a range of other challenges and risks come into focus. Public discussions can take place over, for example, the pros and cons of nuclear power in relation to other energy sources, the nuclear waste issue, the risk of reactor accidents, or the aftermath of disasters that have already occurred, such as in Fukushima and Chernobyl. What often happens in these public debates is that technical experts – for example, experts on nuclear fission – are given a prominent role in discussions over these multifaceted societal risks also – though very seldom is it made explicit that these experts are not experts on these broader kinds of risk. In other words, the limits of scientific expertise – what the expertise is about – are not made explicit. A consequence of this is that public rejection of the tendency of experts to exceed their area of expertise is interpreted by experts and decision makers, as '(mis)judgement of the risks per se' (Wynne 2001: 447).

When such misinterpretations of public judgment forms the basis of efforts to educate and convince a public that is assumed to be ignorant, this leads, according to Wynne, to 'the systematic patronization of the public as intellectually vacuous' and to 'the protection of scientific institutions from the necessary process of critical self-reflexivity about the implicit limitations and contingencies of their own knowledge' (2001: 447).

Instead of admitting that many of the risks and problems associated with nuclear power do not have any clear factual answers, the gap between experts and laypeople, between those who know and those who do not, is reproduced. A small elite of experts, who adhere to a narrow technical framing of the problem, defines what the issue is about and how it should be discussed. Political decisions are also often shaped by these narrow definitions. Wynne discusses these in relation to genetically modified organisms, and earlier we discussed them in relation to the case of nuclear power. With regard to these issues, science would *ideally* be an intellectual resource for policy making and decision-making. However, the tendencies pointed to here are, rather, about supporting an unreflective policy culture and a 'correspondingly unaccountable public representation which this policy-scientific culture imposes' (Wynne 2001: 448).

Irwin and Wynne's point is not to question the self-evident point that experts know more than laypeople about the small area of expertise they specialize in, since that is why we have experts in the first place: experts would not be experts if they lacked access to specialized knowledge. Their point is, rather, that the deficit model is allowed an unduly influence over discussions and conclusions where it does not belong. In such situations, it is assumed that deficit is a problem, and that laypeople must therefore be educated so that the knowledge gap between experts and laypeople is diminished. The assumption is thus that without specialist knowledge, people will make the wrong conclusions around the issue in question. The goal is to make experts and a reasonably scientifically literate public to draw the same conclusions in the light of the same knowledge. The image is clear: the public sits in the dark cave but should come out to be enlightened by science (see Chapter 2).

According to Irwin and Wynne, the deficit model underestimates the knowledge and experiences of laypeople and highly overestimates the role of expert knowledge in discussions and decisions over controversial risk issues that relate to science and technology (see also Irwin and Michael 2003: 22). Wynne (1993) argues that in principle – although not always in practice – there is always room for public knowledge when it comes to decisions about science and technology. These knowledges can enrich expert knowledge and are crucial in making judgments about the relevance, ignorance, and trustworthiness of expert knowledge. For example, laypeople have experiences of judging the trustworthiness of experts and expert knowledge in relation to complex issues. The public should not be seen as empty containers that need to be filled with specialized knowledge in order to be able to speak up and take a stance. The dominance of the deficit model in many situations and political practices will lead to marginalization of the experiences of the public and to alienation instead of public engagement. Experts and laypeople will talk past each other.

Wynne strongly criticizes technocracy, and his solution to the problem is that scientific experts should acknowledge the limitations of their own knowledge. In this way, the deficit model can be overcome and the communication between experts and laypeople improved. To the extent that scientific experts have more power and influence than the public over political decisions concerning science and technology, we should also demand that they reflect on their limitations. Experts have nothing to lose by demonstrating their reflective capacities; on the contrary, this would strengthen their legitimacy. Similar to Callon, Wynne argues that scientific experts risk losing their legitimacy if they maintain and protect sharp divisions between experts and laypeople. This would only heighten conflicts, strengthen the problematic deficit model, and endanger the authority of experts and expert knowledge where public concerns are not responded to, causing a situation of mistrust.

In order to strengthen the legitimacy of science in the political sphere, there is a need to understand scientific results as not completely certain and to recognize the limitations of scientific expertise. This includes acknowledging scientific uncertainties as well as the dimensions and aspects that are excluded from scientific studies and accounts and which it is therefore impossible to say something about with reference to studies that have been carried out. A reflexive science integrates such dimensions in its own knowledge formation and is not afraid of acknowledging alternative understandings of, for example, risk. In decisions about how science and technology should be regulated, there are good possibilities for developing reflexive approaches. The success of the regulation can be evaluated in relation to its capacity to be flexible: 'it may require that some imprecision and ambiguity of formal regulatory standards and definitions be maintained, as an adaptive arena in which the contending parties can interact, negotiate, and settle and renegotiate the practical meanings as they go along' (Wynne 1987: 9, italics removed).

For the social scientist, the task is to study how problems are framed and to make explicit how power relations between various groups are shaped and affect how issues of risk are defined and how knowledge and opinions are categorized and evaluated. Wynne's approach critically targets technocracy and experts' influence over people's lives. He is explicitly normative and argues that risk issues are often narrowly framed, which leads to problems of legitimacy. Moreover, he argues that these narrow framings must be openly acknowledged as simplifications and be discussed in public debate as a democratic principle and 'in order to save scientific culture from itself' (Wynne 2001: 472).

Collins and Evans argue that Wynne is right in his critique of scientific experts, but that it is irresponsible to not maintain any division between experts and laypeople. Wynne becomes, in Collins and Evans model, a representative of wave two, which leads to a romanticization of the public and thus risks becoming allied with populism (see Chapter 3). Thus, the focus is yet again on participation: who is included and has influence over the definition and selection of issues, for whom, and with what consequences. In the next chapter, we expand on the issue of participation.

Politicization of expertise

Mark Brown argues that politicization of expertise is a very common situation today. According to Brown (2009: vii), many commentators agree that this situation is unwanted, and therefore suggest various solutions to avoid politicization. One suggested approach is to get rid of political elements in science and reconstruct science as value free. This includes guarding the boundaries of science through strict peer review and excluding all contributions that do not live up to scientific standards. The scientific domain may thus shrink, but it will be purer and more certain. Separation from politics is a central feature. A radically different approach is to let more voices be heard in the scientific debate – for example, in the debate on climate change. The argument is that scientific results are questioned when there is a lack of legitimacy for science to speak with one voice. There are facts and perspectives that are pushed aside due to the demand for scientific consensus. A suggested remedy to this is to let the public, or groups of the public, discuss and comment on the relevance and role of science, not least its significance when it comes to particular decisions, such as climate change mitigation measures, new medical treatments, or the introduction of genetically modified organisms. Jasanoff (1990) has referred to the two approaches as technocratic and democratic (see Chapter 4).

The two suggestions are radically different. The first focuses on how science should represent its object of study (nature), while the other focuses on how science represents public will. However, an aspect that seems to unify all those who criticize the politicization of science is that they see representation as problematic, and what is preferred is a less mediated representation in which science should become closer to nature and/or public will. Politicization is thus seen as something negative, and representation is seen as both the problem and the solution: a better, more true representation is sought, and it is thought that it can be accomplished if it becomes less mediated. Therefore, mediation is understood as a distortion, a misrepresentation that leads to politicization.

Brown disputes this description on the basis that mediation is something unavoidable. Science *is* political, and this is not something we can or should change. The first approach, he argues, is therefore insufficient or even impossible. The second approach is more promising, but still underdeveloped.

To criticize the politicization of science is in principle wrong as it assumes that science can stand outside and above the political (Brown 2009: 2). From this perspective, which is defended for being purely scientific, other views are criticized for being politically influenced. The aim of eliminating politics from science leads to conflicts over what is political and the risk is that political conflicts take centre stage also in scientific contexts. In short, political conflicts are hidden and understood as scientific conflicts. Brown argues that this in fact implies a politicization of science, since it claims that values and political judgments do not have their place in scientific contexts. A politicization of science and a scientization of politics are, for Brown, two sides of the same coin – that is, politicization. The latter is the dishonest type since it hides political conflicts and make them more difficult to manage. This is the classical aim for separation: the republic of science that says that science can and should regulate itself, and only then can it be useful to society (see Chapter 4). But this wish for separation is both impossible and irresponsible (Brown 2009: 17). Expressed differently, we could say that science is neither political nor apolitical, but it can always be *politicized* – that is, it can be *made* political. In co-productionist terms, we can understand a separation between science and politics as temporary and something that can be challenged by efforts to reveal that issues are hybrid.

Brown's (2009: 185ff) view that science can always be politicized resembles Callon's discussion about 'hot' and 'cold' issues. To take this discussion a step further, Brown argues that our task is to study how and when science becomes politicized, and also when it does not – when it remains cold. This implies the empirical aim of studying what the actors do and how they make judgements over what is political and what is not (Brown 2009: 187). Politics, according to Brown, also has a substantive side, which is about power, the collective action in which conflicts over goals and means are at stake. This substantive definition means that everything can be political but also that not everything is political – that is, when power or conflicts are missing or suppressed (Brown 2009: 188).

When politicization occurs, there are often counter-reactions. This can lead to efforts to move or suppress the conflict, which means that the conflict is not openly admitted as political and that the counter-reactions are directed to eliminate the political from the scientific (Brown 2009: 192). Because of such tendencies, Brown argues, it is only through an *institutionalization* of politicization that a constructive democratization of science can be achieved.

Institutionalization means, according to Brown, that politicization of science must be accepted as a positive potential and not as something negative. Science is always *potentially political*. Processes of politicization can be initiated from within or through external input, and new issues can be introduced in politics as well as in science (Brown 2009: 197). It is, therefore, crucial to not assess consensus, delegation, or trust in experts and conflicts and resistance as either/or issues, but in terms of both/and, and as a dynamic between politicization and depoliticization. A key issue in this context is how power conflicts and illegitimate power relations in science are managed. Brown argues that this depends on whether there are institutions that can support democratic governance to even out power relations. 'Contesting such power relations—that is, politicizing science—need not reduce expert advice to partisan competition, provided that institutions are available that equalize power and facilitate mediation among diverse perspectives and interests' (Brown 2009: 260).

Brown's suggestion can be seen as support for Callon's hybrid forums, and the view that spontaneously emerging hybrid forums are often in need of organized cultivation. Rather than trying to separate science from power, power needs to be distributed more evenly among institutions and between institutions and social groups. To focus on participation and how participation can be organized is then an important aspect, crucial for a politicization of science that can be productive and constructive.

Understanding politicization as something necessary, and avoidable, also means that we need to take an interest in how participation and representation works and whether extended participation gives more, or perhaps less, space to acknowledge and contribute to constructive politicization.

Conclusions

In this chapter, we have presented the idea of co-production as something essential in the STS scholarly tradition, from the early aim of explaining knowledge development as part of social, and group, formation to the idea of mutual dependency and tandem development of knowledge, actors, and social contexts, and Jasanoff's aim of analysing the co-evolution of scientific knowledge and social order on a macro level.

Latour's emphasis of the two (connected) processes of purification and hybridization makes an important amendment to the notion of co-production which helps us to better see the connections between separation and co-production. This means that the many separations that are viewed as having a crucial place in modern society can also be understood and analysed from a co-production perspective, and often these are developed in conflict and in response to hybrid activities. Such double processes in the interplay between science and politics are similar to the processes which Callon refers to as 'hot' and 'cold' and what Göran Sundqvist (2014) calls processes of 'heating up' and 'cooling down'. This means that separation, in an STS analysis, does not need to be understood as completely different from, and opposite to, co-production. On the contrary, separation is another ambition and a different result based on the same kind of co-production processes where science and politics are always entangled.

What is also noteworthy in relation to STS work on co-production and hybridization is a normative ambition to improve the interplay between science and politics. This is most visible in the work of Callon, Wynne, and Collins and Evans. Their approaches are based on the assumption that expert knowledge is often misunderstood and made into something unexplained, but also something that can be used as a norm for assessing types of knowledge which are interpreted as lacking a scientific foundation. Since knowledge development is always about co-production between knowledge and social factors, it is impossible to assume something is autonomous, which leads to attempts to defend something that can't be demonstrated – that is, 'pure' and 'autonomous' knowledge. From this position, STS scholars present an alternative. This means making normative arguments about what knowledge is and how it works, not least in political contexts, based on theoretical assumptions – in short, a theoretically based normativity that we have presented using the work of Callon and Wynne.

This theoretical foundation, based in the idea of co-production, acknowledges variations and alternatives as well as the importance of reflexivity and learning. Here, we find the STS statement that things 'could always be seen otherwise' (Woolgar 2014: 31), which also connects to Brown's focus on what is becoming and processes of politicization. The world – co-produced nature and society – leading to the development of 'assemblages' means a changing world which is always becoming. Knowledge is always co-produced, which means that knowledge and society are always intertwined and inseparable.

Accepting the idea of co-production, however, does not mean that we always need to take a stance against separation. On the contrary, understanding and making co-production explicit means that we have a choice. If certain issues are seen to be better handled by experts and through delegation to technology, this is a result of co-production – that is, a result of the authority, status, and presumably (albeit not always and certainly not by everyone) legitimate mandate given to experts and technological solutions by societal actors. How openly such delegations are discussed, redefined, and tested differs from case to case depending on many aspects, such as the history and policy cultures of the issue in question. Co-production is not always – and, in fact, is very seldom – made explicit. Therefore, STS analyses are of great importance, since these are about making such processes visible. When analysing processes of co-production, we can be much helped by focusing on actors – who the actors are and what they are doing. In short, we should focus on *participation*.

6

Participation as Co-production

Introduction

Both science and democracy are based on representation. We have suggested that this means a dividing line is created between those who represent and those who are represented. Concerning science, and expert knowledge more generally, fears are sometimes raised in relation to expert rule. However, this is often an unproblematic, and even a desirable, situation. The desire to hand over certain tasks to experts is especially strong when non-experts have no interest in knowing the details, or would not even be able to learn them, but still need to get things done. This applies to everything from healthcare to plumbing, architecture, and energy production. But sometimes engagement and resistance can be mobilized among citizens because the representation – the delegation to experts – is unsatisfactory. Experts can be wrong and can claim authority on issues that go far beyond their expertise and yet have great influence over the course of events. Citizens can sometimes contribute to better representation since they have other, often more local, experiences than the experts. But above all, citizens, relevant groups, and the general public have an important role to play in judging expertise as credible and relevant, or not.

Participation and its role for democracy is much more complex than a simple dividing line between unquestioned delegation to experts and direct participation of laypeople. It involves parliamentary debate and decision-making, practices of public policy making and administration, issue formation and mobilization in social movements, practices of investigative journalism and media reporting, and so on. This also means that there are potentially many democratic sites at which science and technology issues are communicated, decided on, and practiced (Laurent 2017). It means too that for each issue, we might be able to trace particular ecologies of participation – for instance, particular forms of communication and dialogue, assemblies of concerned groups, types of facilitator for participatory events, and institutionalized ways of managing issues in inclusive or exclusive ways (Chilvers and Kearnes 2016).

This chapter returns to the idea described in Chapter 4 of participation as crucial for good governance, exploring this in more depth and in relation to the idea of co-production. In short, participation is understood as being about co-production of issues and actors. In the next section, we discuss why STS is particularly concerned with the question of participation. Then we describe the interest of government bodies and other expert organizations in listening to publics. We present what we call *organized participation* – that is, participatory processes organized by someone who invites others to participate and who has certain ideas about why participation is needed and who to invite and why. The fact that some are invited and welcome to participate also means that others are uninvited, perhaps forgotten, and in some cases unwanted. We discuss the distinction between, on one hand, publics who are concerned with the decisions and plans made by governing bodies and, on the other hand, expert organizations and social movement groups who mobilize and engage around specific issues. Then we explore in more detail how participation can be organized and what aspects are combined (co-produced) in a participatory process. The logic of the organizer must somehow match with the invited participants and, not least, with the issues that the participatory process is supposed to focus on. Even if the intention is to invite in order to create a broad and open dialogue between concerned parties, a lot of other things are going on simultaneously. Therefore, we focus on the eventfulness and unpredictability of participatory processes: participants are not always doing what is expected of them, and, in addition, the organizers are not always doing what they say they will. In the final section, we discuss the advantages of studying participation from the approach of co-production. What does this allow us to see, and what can we thereby understand better about the relation between science and democracy?

Why participation?

STS points to representation as a crucial issue and suggests that the remedy to technocracy – which is seen as a too narrow and illegitimate in terms of representation – is to expand and improve citizen participation. This position in favour of participation has been criticized. For example, Brown (2009) argues that STS seems to focus on participation at the expense of democratic institutions. In addition, he shows that representation is much more than participation: it is about the legitimacy of power and accountability as well as the similarity between the represented and the representers. Are opinions being represented, or is there even a similarity between representers and represented? Do representers share the gender, class, and ethnicity characteristics of their voters?

Questions concerning representation, participation, and democracy are well-known in political science. But democracy is only one of two pillars holding up a society based on representation. The relation between participation and representation in relation to science has not been discussed much, other than in a small part of STS research. The way STS understands science from the perspective of representation implies an interest in participation.

STS scholars in support of extended participation have been accused of being indiscriminate and naive (Moore 2010). At the same time, we can also find a critical and almost cynical approach to extended participation among STS scholars. Brian Wynne has been most vigorous in his criticism of the often-found technocratic framing of participatory processes and, more generally, expert communication with publics. When such an expert culture tries to convince a public presumed to be ignorant of risk issues, misrepresentation occurs and leads to a systematic underestimation of the substantive role of the public, and to 'the protection of scientific institutions from the necessary process of self-reflexivity about the implicit limitations and contingencies of their own knowledge which is being given unqualified sovereignty' (Wynne 2001: 447). This critical strand of STS recognizes the ways in which new participatory governance styles often become hijacked by old technocratic governance structures.

It is not easy to find a balance between more and better scientific expertise, and more insight and influence for the general public. On the contrary, there is a longstanding dispute between those who claim that more 'participation' is what is needed and those who argue for 'expert rule'. Jasanoff (1990: 15–19) asserts that this dispute is based on a misunderstanding. Proclaiming a technocratic solution (with a focus on facts) *or* proclaiming a democratic solution (with a focus on values) are based on the similar conceptual mistake and assumption that there is, on the one hand, a stable, objective, value-free science and, on the other hand, stable political and social values.

STS research has shown the value of extending participation when representation is too narrow (Soneryd and Sundqvist 2022). The problem of narrow representation emerges when sharp boundaries are created around the relevant science in relation to a given question and, at the same time, the question under scrutiny potentially concerns many people. Competent authorities do not always perceive a concerned population as capable of contributing with anything substantial in relation to questions concerning science and technology. Wynne (2007: 108) argues that competent authorities often simply misunderstand the role of the public as well as its capabilities. Are demands being made on the public to make contributions to a technical expert debate? Or, rather, is it about public input into a discussion about public issues that *relate to* technical expertise? According to Wynne, the general public is competent in assessing the validity and relevance of ideas that experts express when it comes to their future use, identifying the possibilities and costs/risks of research investments and plans, and, not least, valuing the credibility and legitimacy of experts.

Government authorities, or other expert-based organizations, that invite the public to a dialogue on issues related to science and technology will have ideas about the benefits of the discussions, what qualities the invitees possess, and what the invited participants are expected to contribute.

Based on assumptions that expert knowledge has value and that a technical debate requires a certain exclusivity, ambitions of STS research to democratize processes by increasing participation can be formulated in a specific way. Extended participation is justified when this can improve the representation of issues that are related to technical expertise but which are kept implicit and hidden in expert opinions or decisions based on science. Only when representation can be improved with more participation is that participation justified. But it is still not a given how this idea should be implemented or how participation can lead to better representation. When and for what issues is extended participation a good idea? How should participation be organized? Who should be invited? These are key issues for STS research.

Listening to the public

In previous chapters, we have shown that authorities and other expert-based organizations which rely on a deficit model will have an interest in one-way communication only. Experts are seen as having a role in educating publics, but the reverse is not considered valid. However, it is always of great value for decision makers to listen to publics, and not least to the 'silent majority' – that is, those who are not organized or mobilized around a particular issue (Lezaun and Soneryd 2007). The silent majority can be understood as the greater part of the population that has not yet formed an opinion. The aim of listening to these silent citizens is motivated by the fact that those with an opinion are often loud and tend to drown out others and other perspectives which may be equally legitimate and worth listening to. Those who are already mobilized and have formed an opinion can make themselves heard through protest, activism, or lobbying.

However, since the general public are quiet, decision makers must listen actively. A lot of work is required for silent voices to be heard. There are various tools and methods to accomplish this, such as surveys, focus groups, consultation meetings, and citizen panels. These methods can generate lay opinions in relation to controversial issues concerning science and technology. Thus, the active listening to publics requires method competence. Competences and knowledge about methods – their different rationales and their pros and cons as well as the practical know-how of invitation, moderation, and summarizing results into reports – can be seen to form *participatory technologies* – that is, assemblages of expertise and technical devices, and the implementation of these in practice (Soneryd 2016).

Representative democracy is often seen within the borders of the nation state. Here, relevant publics for decision makers to listen to can be demarcated as the populations within national boundaries or in other political–administrative entities, such as municipalities, or regional entities, such as the EU. Such political-administrative units are relatively stable. However, potentially concerned publics are fluid, with the boundaries constantly in flux as people are mobile; also the consequences of decision-making move across borders – for instance, technical innovations are rapidly diffused transnationally, and environmental impacts and pandemics know no national borders.

Corporations or intergovernmental bodies, who want to relate to relevant publics, can themselves be located in various places at the same time. Here again, the boundaries demarcating relevant publics are therefore always fluid. When organizations decide to listen to publics, boundaries will be drawn for practical reasons. After all, specific groups need to be reached with invitations to take part in meetings, digital forums, or surveys, and decisions need to be made on who to reach and by what means.

This implies that the production of lay opinions will generate new experimental clusters of people. These can be understood and managed through social scientific and psychological expertise. Experts having competence in the methods of participatory technologies have a prerogative when it comes to interpreting the results and can therefore claim to represent the lay opinions that they have organized and measured – that is, the 'new "experts of community"' (Rose 1999: 189). The new entity that is created is a group, or cluster of people, recruited from the silent mass, whose opinions it is now possible to tap into via the answers they have given in a survey or through their dialogue in a focus group meeting or a hearing. Opinions are not 'out there', and they cannot be heard directly; representatives who can produce, interpret, and report on the results are needed. In many countries, activities aiming to report on lay opinions or to stimulate dialogue between experts and publics have become mandatory and routine. As Alan Irwin notes, 'even the most science-centred governmental report is incomplete without a section on "public engagement"' (2006: 300).

That there is value in listening to the public rests on three assumptions. First, laypeople can contribute with important insights, knowledge, and values that can enrich decision-making related to science and technology (Wynne 2001). Second, consultation and participation can remedy public mistrust in new and controversial technologies and, therefore, also uphold the legitimacy of regulating institutions (Wilsdon and Willis 2004). Third, and related to the first two assumptions, silent majorities are more valuable than already mobilized interest groups, since they are easier to move (Lezaun and Soneryd 2007). In the next section, however, we show that variations in who is asked to contribute are great. Sometimes the general public is invited, while at other times it is interest groups and predefined stakeholders that expert and government bodies want to listen to. In both cases of targeted invitees, however, a byproduct appears, which is the *uninvited*.

Invited and uninvited publics

Just as an 'ignorant mass' is created in relation to the two elites in a delegative democracy (see Chapter 2), so can participation be studied relationally. A theme throughout the work of Wynne (2016) is the relational ontological approach and a critique of decision makers and regulatory experts that deny this relational ontology. A relational approach implies a focus on how 'scientific expert knowledge embodies assumptions and commitments of a human kind, about social relationships, behaviour and values' (Wynne 1996: 68).

What we have described earlier as a deficit model (see Chapter 3) is based on the public understanding of science. However, in Wynne's relational ontology, attention is turned towards the scientific experts and their understanding of the public. Participatory processes organized by expert authorities are always undertaken in relation to specific issues and ideas that motivate participation, which also shapes how participation is organized. Wynne (2007) has described how top-down participatory processes are characterized by the difference between the invited and the uninvited.

Sometimes participants are self-selected. For example, a responsible organization might arrange a meeting and announce it publicly, in which case the meeting is open to anyone who sees the invitation. A danger with self-selection is that the recruitment of participants is biased, as those who have already formed an opinion are the ones who will turn up. The organizer might want to counteract self-selection due to the risk of biased recruitment and come up with ideas about how to create a broader representation. The biggest problem with all forms of invited participation is that they are framed in ways that are bound to the organizers' assumptions about what they want to discuss, *not* issues raised by an engaged public. This framing limits participation since it is based on both explicit and implicit assumptions about what is important to discuss and what knowledge is relevant (Wynne 2007: 107). According to Wynne, it is these kinds of assumption that the uninvited public – that is, groups that are engaged and mobilize around plans and decisions – often challenge.

An example of an organized event that created a distinction between invited and uninvited publics is the broad and multifaceted public debate around genetically modified (GM) foods in the UK, initiated in the early 2000s by government authorities responsible for agricultural issues. The idea was that the debate would take part all over the country through self-organized groups, meetings, and lectures. 'GM Nation', as the public debate was called, was motivated by questions around the future of GM crops in the UK. When the debate was underway, the organizers began to see problems that they associated with self-selection. In response to this, they organized a parallel dialogue process in which only people with no preformed opinions or knowledge about GM crops could participate. The self-selected suddenly became the unwelcomed, and a distinction was made between a pure public without any opinions and groups with entrenched opinions (Lezaun and Soneryd 2007).

STS studies of participation rely on the assumption that there is no public 'out there'. Publics are formed according to the ways governing bodies address them and in relation to regulatory failures that might lead to issues that engage publics. Authorities can address publics as passive receivers of information in an educational initiative, or as partners invited to take an active part in dialogue. But authorities can fail to represent important issues or groups of people, and those parts of the public (the uninvited) can then serve to make the governing bodies aware of their own blind spots. The way governing bodies manage the tensions that arise between the publics they have invited and self-selected activists, who remain outside this formal invitation, differs across policy areas (Welsh and Wynne 2013).

Organizing participation

A strong participatory trend, which can almost be described as having global scope (Wynne 2007), has led to public engagement and dialogue initiatives in most policy processes concerning science and technology (Irwin 2006: 300). But there are variations in how such initiatives are realized in different countries and political cultures – depending on the history and framework of institutions, in the form of legislations and regulations – and in different policy areas. STS research on participation has often involved empirical studies, based on in-depth case studies, of how participation is organized and the implications in terms of inclusion and exclusion of issues, perspectives, actors, and material objects (Marres 2012).

There are many advantages of studying how participation is organized. It allows us to ask questions about who is organizing, who decides who is a legitimate participant, how organizing can fail, how organizing creates order and places certain demands on participants, and how participants – invited or uninvited – can resist the orders that the organized participation both presupposes and tries to establish. Attempts to involve the general public can also be based on different motives or logics. Next, we briefly describe three examples that illustrate different contexts and logics for extending participation.

How the EU Water Framework Directive has been applied

Our first example is from research on how the EU Water Framework Directive has been applied. The framework directive encourages the involvement of relevant groups in water management (European Commission 2000, 2003) and has been described as 'one of the most important and most ambitious pieces of legislation in the history of the European Union's ... environmental policy' (Bourblanc et al 2013: 1449). Framework directives give relatively large freedoms for EU member states to formulate how the requirements of the directive are to be applied in national legislation. The framework encourages stakeholders – those who use and influence water, such as agriculture and industries located by watercourses – to be involved in water management. Participation is motivated in this case both substantively and instrumentally. The goal of better water quality is seen as being achievable only with the involvement of those who can contribute to better water management (instrumental motive), due to their knowledge about local water quality as well as their behaviour affecting water quality (substantive motive).

Water issues are genuinely transboundary and cannot be managed where there are sharp boundaries between academic disciplines or between scientific knowledge and local experienced-based knowledge (Prützer and Soneryd 2016). Nevertheless, water management is characterized as being highly dominated by natural science. The ambition to extend participation to water management rests on the instrumental idea that interest groups should come to an agreement. A more productive approach to knowledge, however, is to accept conflict and strive for critical but respectful questioning of both traditional expertise and experienced-based local knowledge. Transboundary knowledge and critical approaches are needed to avoid lock-ins and problems being concealed rather than attended to. At the same time, Tobias Krueger et al (2016) argue that it is important to realize that an open attitude towards knowledge cannot be forced on local actors. For example, in one case they analysed, participants from the local population expressed a preference for a more traditional division of labour: as a reaction to efforts to involve local people in hydrological modelling, the local population expressed the opinion that the researchers should make the decisions, since they know best (Krueger et al 2016). It is not always the case that groups in the general public demand participation or more influence, but the responsible organizers must offer opportunities for those groups to contribute with knowledge if conflicts emerge. To assess water quality and formulate goals for improving it, it is necessary to involve the people and enterprises that use the water. When water users do become engaged in water management, they are part of already formed networks of experts, hydrological models, standardized definitions, and reference values, all of which may make the process less open to the input of local concerns.

Public engagement on GM crops

The second example concerns GM crops and is different from the first since it is not about resource management, but about innovations that are diffused in a global market. The problem in this case is not related to environmental impact and risk associated with GM crops, though this is an important focus for responsible authorities and researchers. These aspects are regulated and assessed before crops can be grown on a commercial scale. The problem that has been the focus for expert authorities is public worry and mistrust of GM crops. Many efforts to involve the public in dialogue with experts on biotechnology or GM crops have been motivated by the wish to create trust in the new technology as well as in regulating authorities.

Top-down initiatives to engage with the public, which draw sharp lines between 'publics' and 'stakeholders', have dominated when it comes to dealing with public concerns about emerging technologies (McNeil and Haran 2013; Reynolds 2013). For issues concerning biotechnology, particularly issues concerning GM crops, a common way of initiating and organizing dialogue between experts and laypeople is through 'consensus conferences' (Seifert 2006). The consensus conference is based on the idea of inviting a cross-section of the population, with the condition that these participants do *not* have specialized knowledge and are not already part of a mobilized engagement around the issue under discussion. That the consensus conference became a frequent method for engaging with publics, in particular in relation to biotechnology, is not due to any perfect match between method and issues, however. The 'success' of this specific method can, rather, be explained by the networks that support and use it (Soneryd and Amelung 2016).

The climate issue and approaches targeting consumption

A third example is the climate issue and approaches targeting consumption, in which there is not much room given to a wider public to contribute to debate and decisions since engagement of the public is limited to people's own climate-friendly consumption choices. Climate change is a good example of an issue that is framed as an expert issue, dependent on science, but has also frequently been framed as an issue of individualized responsibility. The latter is related to modes of governing through market actors and market mechanisms. A consequence of governing through markets is that in climate campaigns by authorities as well as in public debate and popular culture, citizens are approached as consumers (Dahl 2014; Soneryd and Uggla 2015). Green consumerism, however, is primarily symbolic and tends to focus on particular products rather than on individuals' entire consumption patterns. Thus, green consumerism might allow people to act as individualized consumers and responsible citizens at the same time, but without substantial reduction of consumption (Blühdorn and Deflorian 2019). For example, exchanging your petrol car for an electric one or installing solar cells on the roof of your house can be such a symbolic act, which also may be subsidized by governments through tax reductions for green technology. Compensating for the impact of your flight travel through paying for carbon offsets is another example.

The conditions for citizen participation

Within these three broad issue areas, we find different contexts and conditions for citizen participation due to legislation, modes of governance, dominating expertise, and issue framing. The governance of water management, genetically modified organisms, and climate change is composed of very different actor networks, or ecologies of participation (Chilvers and Kearnes 2016), in which ordinary citizens have different roles and may find it more or less difficult to participate. The networks of expertise, technologies, and modes of governance that surround various issue areas can be more or less governed from the top and thereby more or less open to inclusion and participation.

The conditions for citizen participation can thus be understood in relation to complex compositions of regulation, expertise, framings, and forms for inviting a part of the general public or engaging stakeholders or consumers. These compositions assume and establish particular orders; however, there will always be room for resistance. Not even the most stable framing or structured organization of participants can decide how invited or uninvited publics will react, or demand that they will accept the framing in question.

Jasanoff emphasizes that the social and political context will affect assessments of what is 'good knowledge' according to what is politically useful or culturally preferable. Jasanoff uses the notion of *civic epistemologies* to capture this. The public meaning-making around science, technology, and expertise includes how both decision makers and concerned publics relate to national and international regulatory frameworks (Jasanoff 2005: 255). Inevitably, nation states have a particular role, since the possibilities for citizens to have insights and influence in public decision-making is affected by the nation states' ways of organizing and managing particular issues. Through these ways of organizing and managing, a political and civic culture is shaped, which will have continuity but will also vary over time and differs between nations.

In a study of three technoscientific issues – nuclear power, GM crops, and nanotechnology – Ulrike Felt (2015) shows that the meaning-making and resistance against these technologies among Austrian citizens can be understood as grounded in the country's strongly anchored idea of national political freedom. This idea goes back to 1955, when Austria became independent from the post-war alliance that had governed the country after World War II. Citizens have activated this identity in relation to these three technologies, which are understood as innovations created outside the country, mobilizing against the introduction of the technologies with the aim of keeping Austria free of them. This is an example of how a nationally anchored civic epistemology can relate to and affect the future of technical innovations.

When more actors are invited into participatory processes, this means that people, ideas, and perspectives are combined in new ways. By exploring which elements are combined in participatory processes, the question of how relevant concerned groups and legitimate knowledge claims are co-created in specific contexts can be studied (cf Wynne's relational ontology, mentioned earlier).

A few decades ago, it was enough for government bodies to invite people to take part in citizen dialogue through advertisements in the local newspaper, and to offer coffee and a sandwich while technical and scientific experts presented their plans to establish enterprises that might have environmental impacts. Today, there are greater demands when it comes to organizing citizen dialogue in a professional way. In a participatory event governed from the top down, several things need to be combined in an active and reflexive way: the methods, the selection criteria, the moderation of the dialogue, the participant interaction, and perhaps the hiring of consultants or moderators who are experienced in the selected methods. There is a range of possible methods to choose between – citizen panels, focus groups, hearings, and consensus conferences – all with specific criteria for who to invite and how (Ackerman and Fishkin 2004). Different methods are associated with specific ideas about how participants should interact with each other and what the organizer wants to elicit from the dialogue, which might require specific expertise and moderation. Sometimes extra work and resources are required to collect and present information and stimulus material to be handed out before the dialogue. The more standardized the technology for inviting and engaging with citizens, the more predictable the results. At the same time, participation is always inherently unpredictable, since overflows might appear – that is, there is always a risk that unforeseen effects or unrepresented phenomena will arise, which would call for attention.

The unpredictability of participation

Every society is organized, and often there are clear boundaries between activities and fields even though the world consists of constantly emerging hybrids. Extended participation, initiated with the aim of opening up perspectives that are not yet represented, is a response to existing and established boundaries. Yet, the initiators of events or processes that aim to acknowledge these responses might have narrow and entrenched ideas about how the new perspectives are to be categorized and managed. The unpredictability of participation means that the result is not always what the organizer had in mind. Participants may not behave as expected, uninvited groups may appear, and overflows that alert to the insufficiencies of the expert framings may arise.

It is easier perhaps to imagine that it is the uninvited and unrepresented who create the most unpredictable events, but invited participants also make sense of the events and may behave in ways that are neither predictable nor controllable by the organizer. Ulrike Felt and Max Fochler (2010) suggest that initiatives to invite the public for dialogue on science and technology should be understood as arenas available to citizens for experimenting and testing their role in society and their position towards the technoscientific developments that are subject to the dialogue. Participants are likely to *both* accept and resist the expert frames that shape the dialogue. Felt and Fochler have studied participants' meaning-making and subtle resistance against the frames within which dialogue takes place, and they agree with Wynne that invited participants tend to accept the given expert framing of events more easily than their uninvited fellow citizens do. Rarely do invited participants act as if they could represent or mobilize a counter-expertise based on their own experiences as, for instance, members of patient groups or local residents.

Nevertheless, Felt and Fochler argue that these invited fora for dialogue still have value, especially if initiators and organizers accept them as unpredictable experiments in terms of people's identities and relations to technoscience. How participants perceive and make sense of their role and participation cannot be fixed beforehand, and the resistance and challenges they may express in relation to ready-made frames and assumptions presented by experts should be seen as 'a creative resource rather than as a problem' (Felt and Fochler 2010: 236). This also suggests that decision makers must be prepared to engage in a more open process and let themselves be affected by the invited participants, rather than seeing the dialogue as part of 'machineries for manufacturing consensus on what have already been labelled "necessary innovations"' (Felt and Fochler 2010: 236).

We have already mentioned that the local events and debates that arose around the issue of GM crops in the UK were not seen as representative, since those who participated were primarily those already expressing strong opinions *against* the issue. As a response to this, the organizer created dialogue groups that invited a 'pure' and 'opinion-free' public through selection criteria that prevented already engaged citizens from participating. However, this did not mean that participants in these groups only expressed positive attitudes towards GM crops. In both the self-organized debates and the more top-down focus group dialogues, there were expressions of perspectives, arguments, and approaches towards GM crops that challenged expert framings. Professionalization of participation and preformulated frames for dialogue do not have to mean that the content of the dialogue becomes restricted (though it can be, and often is), but rather that the status of various statements elicited from public dialogue is dependent on experts in participation. Focus groups, for example, are moderated with the purpose of generating discussions that are as close as possible to everyday natural talk. The professionals that the organizer employs can not only shape discussions in particular ways through the choice of different methods, but also direct interpretation of the results and the context in which these are interpreted. The outcome of the GM Nation public debate was interpreted by the responsible authority through the lens of a sharp separation between scientific expertise and public perspectives and critique. This separation was made explicit when three reports were presented after GM Nation ended: a scientific report, an economic report, and a report on the public debate. The arguments and positions expressed in the public dialogue were not allowed to affect the scientific conclusions, or the economic ones, but were presented separately, as if they had nothing to do with the other aspects.

Another outcome can be seen in relation to the dialogue that the Swedish Radiation Protection Authority (Strålsäkerhetsmyndigheten – SSM) initiated at the beginning of the 2000s between proponents and opponents of a new infrastructure for mobile telephony (see also Chapter 5) (Lezaun and Soneryd 2007; Soneryd 2007). The dialogue took place in several workshops, each lasting one or two days, and were filled with presentations and discussions between experts from the SSM, researchers, activists, and groups of electrosensitives. The composition of these fora was of a hybrid character. For instance, experts from two rather different positions met and had discussions; these were: on the one hand, experts whose research was seen to be credible and therefore constituted the knowledge base for the regulations formulated by the SSM; and on the other hand, researchers whose results were not yet replicated and therefore not assessed as credible. Different hybrid compositions, in terms of groups and technologies, were enacted among those who worked in the mobile phone industry and those who considered themselves ill because of the electromagnetic fields generated from the mobile phones, with these groups meeting in rooms *without* electricity and after all participants had put aside their phones before entering the room.

In this case, even though the SSM invited concerned groups that explicitly challenged the expert frameworks, the experts expressed some disappointment that the participants were not representing the silent public. However, this dialogue did not end with a final report but was continued in a broader scientific council, including representatives who were activists and people with hypersensitivity to electricity (Casula Vifell and Soneryd 2012). To some extent, this led the experts to reflect on their own limitations – for example, by changing the way they talked about the issues. Previously their talk expressed *certainty* around the safety of electromagnetic fields, but they shifted to also recognize some uncertainty. This might be in line with what Wynne is asking for: a more reflexive expertise.

Reflexivity

The argument for a more reflexive expertise, as presented by Wynne and as we described previously (Chapter 5), is that experts need to recognize their own limitations, but also they need to recognize the 'others'. This means drawing attention to a wider public, who should be recognized as being capable of making sense of and assessing technoscientific innovations based on the perceived benefits and risks. In addition, this citizenry may have legitimate opinions about the driving forces for technological development – who benefits from it and who will bear the costs, as well as the ethical aspects and the credibility and trustworthiness of regulatory bodies.

But there is also another meaning of reflexivity which, instead of opening up processes to more perspectives and opportunities, is about constantly trying to improve the mode of governing (Soneryd 2016). Reflexivity is what makes organizations aware of organizational trends and, thus, inclined to adopt organizational recipes and/or standardized policy instruments used by other organizations. The dissemination of organizational concepts and fashions does not happen through some mysterious power, but through powerful organizational carriers who formulate and disseminate 'blueprints' for what counts as good governance. Participation as a trend can be understood in relation to this institutional dynamic and the associated meaning of reflexivity.

Organizations that prove capable of taking public concerns seriously are to some extent reflexive in Wynne's opinion, but they can also engage in separating activities – that is, initiating participatory processes and at the same time engaging in their ordinary activities, without letting these two parallel activities affect each other. The fact that organizations can become better at reflecting on how they relate to the public and more professional at staging participatory processes, without letting this affect other activities, is reflected by the increased demand for evaluations and quality assurance. These activities do not necessarily improve quality, but they make organizations better at formulating quality assurance systems (Strathern 2006: 192). From a democratic point of view, it matters what organizations are reflexive about and how this reflexivity can change previously entrenched views on the issue in question, the relevant expertise, and who is affected and should potentially be involved (Wynne 2016).

Conclusions

A paradox expressed by Andy Stirling is that extended participation is appreciated more in a time 'when processes of corporate concentration, institutional harmonization, and economic globalization render the governance of science and technology ever more obscure and inaccessible' (2008: 263). This means that the link between extended participation and *democratization* cannot be taken for granted. Extended participation leads to new constellations between issues, technologies, and people. Co-production can show the *power* of organizing and give possibilities for analysing how these new constellations are made and how *boundaries* are drawn between the invited and the uninvited and between legitimate and illegitimate issues. This enables us to see how issues and actors are not independent from each other, but are shaped simultaneously.

Co-production always happens in historical and political context. We can explore tensions between, on the one hand, an inviting and inclusive mode of policy culture in which decision makers welcome extended participation and, on the other hand, tendencies of surveillance over troublesome citizens who participate in protest actions and the like. We can explore how such practices are rooted in specific organizations and their context and history.

Studies of participation as co-production show that participatory processes are eventful and unpredictable. An invited participant can have radically different ideas about what it means to participate compared to the organizer; a participant may refuse to adapt to a given order, and the organizer may need to adapt to participants who do not behave as expected.

Studying participation as co-production requires a relational focus. Invited and uninvited participants are created in relation to each other, and through authorities' or decision makers' responses to or neglect of the engagement of citizens and social movements. A relational approach further means that we turn attention to the scientific experts and governing bodies and their understandings of the public, their motivations for when and how extended participation is needed, and how participatory events are accomplished through organizational work and boundary-making. The other side of this relationship can be studied by focusing on both invited and uninvited publics, their responses to the demarcations made by governing elites, the issues that uninvited publics mobilize around, and how these differ from the issues of interest in invited participation. How governing bodies respond to the concerns of uninvited issue publics is yet another relevant analytical focus.

The work of organizing public participatory events involves framing issues in particular ways and deciding on the format for interaction; even high ambitions for openness and public availability can create processes that are open to some issues but not others, and open to some people, groups, and organizations but not others.

The relational approach can also be used to understand the ways governing bodies listen to publics. This need not always be understood as extended participation that gathers more people around the same table or invites a broad set of actors to discuss issues of concern. Listening may also be conducted in the form of surveys that are clearly separated from technical decision-making, focusing only on individual attitudes.

A relational approach to participation highlights the dynamic interplay between efforts to govern and efforts of concerned publics to make their assessments of failures and deficits in current government heard. For such a dynamic to be recognized, we are helped by distinctions between invited and uninvited publics, but also by aspects such as the professionalization of participation, digital divides, and a new focus on open innovation and 'users', which might replace a focus on 'publics' and 'citizens'. The next chapter continues to develop this view on a dynamic interplay between governing elites and concerned publics by introducing a discussion about *scientific citizenship*.

7

Scientific Citizenship

Introduction

Citizenship is formally and historically connected to the nation state, even though this has not always been the case (cf Athens' city-centred democracy). In today's understanding of democracy, however, scientific knowledge and technical expertise intersect with citizens' ability to hold state power accountable to democratic values. Sheila Jasanoff (2017) describes how the nation state concentrates not only political power but also resources that enable investments in 'big science' projects. Based on these concentrations, she asks: if the *demos* should have a role in the technical framing and resolving of public problems, what analytical resources does STS provide to facilitate such participation? The answer to this question, as suggested by Jasanoff, is that ideas and practices around science and technology – priorities, investments, distribution channels, regulations, and so on – are co-produced with ideas about concerned citizens. Thus, STS scholars should call attention to the fact that practices of collective knowledge-making shape our very understanding of the *demos* to be served by democracy.

As we saw in the previous chapter, studies undertaken by STS scholars have focused on the relationships between modes of governing and how issues are made public and open to wider debate, how groups of the public are demarcated through the notion of invited and uninvited publics, and how the agency ascribed to invited publics tends to be circumscribed by instrumental motives. In this work, STS scholars frequently touch on one of the unresolved problems in democratic theory. This is about what properly constitutes *the people*. Yet, any democratic theory is based on an understanding that there *is* a people, a citizenry that is implicated in governing in indirect or direct ways and can hold government accountable. We discussed such unexplored assumptions in the first chapter of this book, and we referred to these as belonging to a shadow theory of democracy. The shadow is an inevitable companion to unacknowledged contradictions and taken-for-granted assumption within established ideas about democracy. If denied, it is a potential threat, but when openly acknowledged and dealt with, it is a potential resource to democracy.

STS scholars have suggested that citizen engagement with and contestation over science and technology should be understood as having a constitutive role in the development of alternative imaginaries of democracy. Such ideas point to the relations between governing bodies and concerned publics that are transgressing the boundaries of the nation state. With Noortje Marres' (2007) conceptualization of publics as emerging in relation to how issues are governed, the citizenry becomes a flexible entity rather than a fixed one. Departing from Jasanoff's question, mentioned earlier, we discuss in this chapter the notion of *scientific citizenship*, as it has been approached in STS studies from a range of policy areas in which science and technology have been contested. We then discuss the benefits of drawing on a conceptualization of citizenship that takes into consideration its transformative potential and dimensions of contestation that often are the focus in STS, as well as its still highly prevalent connection (albeit manifested in various ways) to the nation state. This is carried out with the help of three detailed examples of how scientific citizenship is made in practice; these cover the issues of air quality, user-driven innovation, and food advice.

STS and scientific citizenship

Citizenship is not solely a question of formal membership and the rights and obligations that follow a nation state's constitution, but also something that is actively practised (Leach et al 2005; Isin 2009; Laurent 2017). This view of citizenship is accompanied by an understanding of conflicts around science and the engagement of publics in such conflicts 'as not only epistemic conflicts between ways of knowing, but as reflections of different ways of being' (Leach et al 2005: 5). Conflicts around science and technology can be about endangered ways of living and include struggles over people's health and the environment in relation to tacit knowledge and local practices. The co-constitutive dimension of science and politics can be seen in the ways that genetically modified foods, nanotechnology, and similar heterogeneous applications of science in a market context have been framed as matters of individual concern. Public conflicts around science and technology, as many studies have shown, are contestations of such narrow framings and, thereby, lead to efforts to bring in alternative framings, which point to political, ethical, and cultural issues narrowed down by established actors, such as industry and regulatory government bodies.

STS researchers have used the term 'scientific citizenship' descriptively for analysing what roles citizens are given in relation to issues that are dependent on scientific expertise (Irwin 2001; Elam and Bertilsson 2003; Rose and Novas 2004; Horst 2007; Árnason 2012). In this research, we find studies of the growing interest among government authorities and knowledge-intensive organizations in engaging with publics. In these activities, 'they implicitly define the meaning or generate the practice of the citizen in relation to technoscience' (Goven 2006: 566). When the possibilities for citizens to make themselves heard are circumscribed, so that crucial political choices cannot even be contested, we should ask what a scientific citizenship means. We argue that citizenship must be seen as emerging in a relationship between governing bodies and concerned publics and that this potentially involves conflict. According to this argument, it is precisely the analytical approach of STS scholars that is useful. This means a search for overflows – problems that transgress the initial framings of governing bodies and that can be attended to by concerned publics – which makes STS a resource in an ongoing project of redefining citizenship and democratic experimentation (Marres 2007; Laurent 2017; Soneryd and Sundqvist 2022).

In parallel to studies focusing on how governing bodies approach and appeal to citizens – sometimes in narrow ways, as consumers, responsibilized individuals, or passive citizens entrusting governing bodies to take care of things for them – there are also studies focusing on how citizens mobilize around issues to contest established forms of governance, and sometimes also to establish alternatives in the absence of democratically accountable governance. For example, studies using the term 'biological citizenship' give emphasis to the capability of citizens to form new collectives and use biological knowledge for informed self-governance (Rabinow 1992; Rose and Novas 2004). An example of this is people suffering from health problems due to environmental pollution, such as the situation after the Chernobyl disaster (Petryna 2002; see also Ureta et al 2019). The assembling of biological knowledge about exposure and health effects are, in such cases, crucial for the victims to claim their rights for compensation and achieve public recognition of the harm done.

In the book *Citizen Science*, Alan Irwin (1995) argues that the role of citizens in scientific knowledge production needs to be understood from both sides of the relationship – a science *for* citizens (that is, a science that assists the needs and concerns of citizens) and a science *by* citizens – and notes the importance of recognizing the contextual knowledges which are generated outside of formal scientific institutions for assessing both sides of the relationship. Irwin explores a range of situations involving controversies emerging in connection to governing bodies' failures or deficits, as well as the different ways in which citizens engage with science.

However, for the past couple of decades, the term 'citizen science' has been used more to denote particular forms of the participation of non-professionally trained people in the core activities of producing scientific knowledge. This has been facilitated by digital media and open-source databases that broaden the scope of research and the number of people involved in data collection. The threshold for citizens being able to contribute to scientific knowledge has been lowered. However, the forms of involvement in citizen science projects differ and, with this, the status of the citizen scientist vis-à-vis the researcher. Citizen science projects thus range from more limited and passive citizen roles, which can even be understood as low-qualified, unpaid jobs, to allowing a much more active role and even crediting individual citizens for the discoveries they make (Kasperowski and Hagen 2019).

The actual role of citizens in citizen science is picked up by Barbara Prainsack (2014) when elaborating on criteria for assessing the overall success of a citizen science project. Two criteria are presented: one relates to knowledge contribution and one relates to how well the project educates citizens. The knowledge criterion is about whether the project provides a solution to a pressing scientific issue – that is, whether it *advances the status of knowledge* according to standards and references of the traditional scientific system. While both criteria can be legitimate measures of how well particular citizen science initiatives meet their goals, they might still be narrow measures from a democratic point of view. What is missing from the criteria suggested by Prainsack is the question of framing: what kind of problems is science supposed to solve, for whom, and who has been involved in formulating these needs or problematizations? For example, what should citizens learn through education and why? Sarah Davies and Maja Horst suggest that we need to explore citizenship in relation to both government- and industry-led citizen involvement in scientific institutions or biotech companies, and citizen-led initiatives that imply 'research carried out by social movement and civil society groups' (2016: 192). Their approach is broader than Prainsack's as it includes the wider issues related to framing as well as power relations and societal implications of engaging the citizen in technoscientific issues.

Much of the STS literature on scientific citizenship has focused on how governing bodies draw boundaries between relevant expertise and publics, which define what issues are relevant and what publics ought to know or do. Because of the limited role often given to citizens by governing bodies, a critique has been formulated in relation to studies of scientific citizenship, questioning the use of the term 'citizen' if it is only a passive affair. For example, research in the life sciences has focused on how new biological knowledge has shaped the way we understand ourselves in terms of our genetic and biological constitutions. This allows for new forms of identity taking shape, which has been termed 'genetic citizenship' or 'biological citizenship' (Rose and Novas 2004). Sometimes the label 'citizenship' is used for processes of subjectivation, and in studies of knowledge and governing in a Foucauldian tradition, identities and subjectivities are primarily understood as a passive affair and a result of framing and efforts to govern. The meaning of 'citizen', then, risks losing its distinctive dimensions. Thus, it is argued that we need a vision of 'citizens who are motivated, informed and able to critically reflect on their society and willing to participate in processes of public deliberation about matters of common interest' (Árnason 2012: 930). We believe this critique can sometimes be justified, although we also think that a one-sidedness on behalf of the actively involved citizens and their motivations and imaginaries is also problematic. Focusing solely on the suppressive side of citizenship or solely on the empowering aspects of citizenship provide their own blind spots.

Gwendolyn Blue and Jennifer Medlock (2014: 563) argue that scientific citizenship offers a remedy for such theoretical blind spots, which see citizenship either as a vague and positively connected term or as a means to suppress concerned groups into already defined meanings. Rather, scientific citizenship, they argue, is about 'positioning citizenship in relation to the epistemic dimensions of political engagement and identity, which is to say that it focuses on the means through which knowledge, values, concerns, and interests are articulated and mobilized in the setting of public priorities, agendas, and decisions' (Blue and Medlock 2014: 563). Drawing on Engin Isin and Bryan Turner (2002), they argue:

> Scientific citizenship is grounded in an understanding of citizenship as a dynamic and provisional feature of social and political identity. Similar to other contemporary configurations of citizenship such as cultural, sexual, ecological, technological, and cosmopolitan, scientific citizenship is intimately entangled with the processes, practices, and institutions that enroll people as political subjects at particular times and in particular places. [...] These expanded understandings and practices of citizenship have been spurred, in part, by myriad economic, political, and cultural transformations over the past few decades that have led to a repositioning and reconceptualization of political agency and subjectivity beyond the nation-state. (Blue and Medlock 2014: 563)

Empirically Blue and Medlock (2014) have explored how dominant accounts of climate change animate certain forms of scientific citizenship. Looking at a global public deliberation process aiming to involve citizens across the world in discussions on issues related to climate change, known as World Wide Views (wwviews.org), they conclude that privileged expert-based issue framings dominate these events and also contribute to projecting a passive rather than active form of scientific citizenship for its participants. The dominant framing of climate change, focusing on the global average increase in temperature and carbon dioxide emissions, implies that public discussion is narrowed down to issues about climate change policy and the control of greenhouse gas emissions through technical fixes and market-based mechanisms. These framings not only limit the ways in which people can express their sense of citizenship, but also spur on the ways in which people can voice their scepticism and contribute to a deeper politicization of the process. Other studies have also illustrated this tension between an individualized view of the citizen as a consumer, and a responsible market actor, and a part of a collective political sphere that can question but also contribute to a social, political, and ethical agenda (Vihersalo 2017).

There are also studies focusing on active engagement among citizens and its consequences – for example, through a focus on energy citizenship and the material participation that new digital devices, such as smart meters, enables. In such studies, scholars are exploring

> possibilities of expanding the understanding of energy citizenship and including new publics into discursively and practically taking part in the low-carbon energy transition. We recognize, though, the limits of a materialized energy citizenship, especially in the risk that it creates inclusion of the already advantaged, those with purchasing power and means to acquire new technologies, while excluding others. (Ryghaug 2018: 298)

We believe that the risk that some groups are excluded is rather an integrative dimension of citizenship and something that characterizes the whole history of citizenship, and will continue to do so in the future. Citizenship is a category that draws boundaries between those who are included and excluded, and therefore citizenship is always characterized by struggles over its boundaries.

Studies of scientific citizenship show how citizens' role in knowledge production varies: being passive subjects in education and information campaigns; being participants in various forms of dialogue with experts; being co-actors in knowledge production in citizen science initiatives; and being counter-expertise based on their own experiences (Soneryd and Sundqvist 2019). STS scholars have focused both on the variation in how governing bodies approach citizens and how citizens mobilize around issues. However, there is not always an explicit discussion about the various elements that constitute citizenship.

Citizenship, citizens, and acts of citizenship

Citizenship is a category that both excludes and includes, a category that is filled with examples of both domination and empowerment. The historical connection to the nation state and the understanding of citizenship in its formal dimension, as membership and connected to given rights and obligations within the context of the nation state, is only one understanding of citizenship and also a limited one. When the boundaries of the nation state shift due to globalization and when environmental concerns transgress political–administrative boundaries, new struggles and new sites emerge. How subjects act to become citizens and claim citizenship has substantially changed, and so must our conceptualization of citizenship (Isin 2009). It might thus be reasonable to distinguish between *citizenship*, *citizens*, and *acts of citizenship*.

Citizenship is a relationship that always involves the nation state's efforts to demarcate who belongs and who does not. These boundaries imply a relationship that is both repressive and caring. They mean a stable and long-term relationship, connected to the constitution of the nation state and its rights and obligations. In democratic welfare nations, obligations and rights relate, for instance, to individual freedom, to health and safety, and to participation in collective decision-making. This includes taking part in technoscience matters that the state might be involved in through ownership or investments and as regulator or decision maker, but also those that the state has chosen to refrain from – that is, areas that are not regulated and demarcated as not belonging to the state.

Citizens are those who are formally recognized by a nation state as belonging to the citizenry, but the term also refers to the fact that subjects can recognize themselves as being included in a collective – sometimes this recognition transgresses the nation state, as people might also think of themselves as cosmopolitan citizens or European citizens. This means that when we take on the role as a citizen, another identity is activated, in contrast to, for instance, the role of consumer, employee, or parent.

Finally, there are *acts of citizenship*, which in principle can be performed by anyone – individuals who enjoy formal citizenship and those who do not, those with rights and those without – as long as you act *as* a citizen. To perform acts of citizenship thus means acting *as if* you have the right to claim rights. Acts of citizenship deliberately try to transform the boundaries that demarcate which issues and subjects are within or outside the responsibilities of the state.

Exploring the contours of scientific citizenship: three examples

The examples presented in this section are based on case studies that relate to different policy areas and initiatives to engage citizens. In various ways, they give attention to certain aspects of what scientific citizenship means. The first example is about air quality, and it displays three kinds of initiative that vary in terms of how citizens are engaged: as passive recipients of information, as contributors to the gathering of data, and as being involved in contestation of current forms of government and the shaping of counter-expertise. The second example involves do-it-yourself drug production and is about citizen-driven practices where the state tries to stay at a distance. The third example is about food and dietary advice, and this is about government-led one-way communication to the entire population. All three examples are from an edited volume on scientific citizenship (Soneryd and Sundqvist 2019).

Air quality and the sensing citizen

According to the World Health Organization (2022), close to seven million people die each year because of polluted air. Cities are especially exposed since the pollution from industries and transportation is highly concentrated and affects many people. Air quality control is an area that gathers government agencies expertise from several areas, such as legislation, administration, laboratories, and specialists in measuring practices, including those involved in surveillance of these measures. Data concerning air quality is provided by public authorities through expensive instruments that are located at various sites and have sensors that register the presence of pollutants. In addition, in Europe, measures are based on common standards within the EU to ensure comparability of data from different member states. A sharp boundary between experts and citizens is thus drawn in several ways.

However, today, there are also inexpensive measuring instruments available, which allows for alternative ways of gathering data on pollutants. Digitalization has led to an explosion of alternative ways to measure things, based on sensors that are accessible to many people through their smartphones (McGrath and Scanail 2013). This makes a more active and engaged role for citizens possible. These instruments allow citizens to become part of existing measurement practices. In what ways does this open up new forms of scientific citizenship in relation to air pollution? How are the boundaries drawn around citizen involvement? Why should citizens be involved and how? An analysis by Jesper Petersson (2019) distinguishes between three types of scientific citizenship in relation to measuring air quality.

The first type is based on a science *for* citizens. The government authorities' instruments aim to provide information about air quality for the safety of citizens, but they also become a means to guide citizens' actions to support efforts to mitigate pollutants – for example, by recommending that people cycle on a day when there is a high level of pollutants, rather than use more polluting modes of transport. In this case, the focus is on how the City of Gothenburg, the second-largest city in Sweden, has worked with measures of and information about air quality with the aim of making citizens aware of and getting them to take responsibility for air quality through their own actions, for instance by choice of transport. A science for citizens, in this case, assumes an expert monopoly on the production of air quality data.

The role of citizens is limited to receiving information and acting on this information by making responsible choices.

A second type of scientific citizenship is based on the authorities doing measurements together with citizens. The sites of the City of Gothenburg's measuring instruments are few and fixed, while people with sensors in their smartphones are many and mobile. Because of this, the City of Gothenburg took initiatives to gather air quality data together with its citizens. With the help of citizens, the hope was to get a more accurate picture of the air quality across the whole of the city. This is often called 'citizen sensing', a practice close to many of the activities performed under the label of 'citizen science' (Gabrys 2016; see also Kasperowksi and Hillman 2018). Instruments were made available to citizens, who carried these sensors wherever they went, thus enabling measurement of air quality at many locations.

A third type of scientific citizenship is based on science *by* citizens. Grassroot initiatives to measure air quality are mostly rooted in a mistrust of the data presented by authorities. A citizen group mobilizing around air quality arose during the 1980s in Gothenburg because of heavy pollution of the city by industry. Mobilized groups engaged in knowledge production by creating their own instruments for measuring air quality as a corrective to the exact but inadequate measurement performed by the authorities. The authorities' measures were exact since the instruments were very precise, but they were inadequate since they were only located at a few places. The citizen group not only carried out their own measurement of air pollutants, independent of the authorities, but also made their own analyses and conclusions about the air quality problem.

In this case, use of sensor-based instruments created an interface not only between certain pollutants and nature (as detectors of air pollutants) but also between experts and citizens in terms of who is responsible for data collection, compilation, and communication, and therefore this is an example of co-production of both nature and social order.

User-driven innovation is the future

A subculture of do-it-yourself has emerged as a trend at the margins of the scientific community (Delfanti 2013). Such activities are cherished by heterogeneous actors from the European Commission to businesses, academics, hackers, and home occupiers, all of whom see user-driven innovation as the future. When user-driven innovation is discussed, most often harmless examples are given, but inordinate optimism about user-driven innovation can be deeply problematic. A case study by Johan Söderberg (2019) focuses on illegal user-driven innovation – more specifically, the extraction of dimetyltryptamin (DMT). DMT is classified as a narcotic substance, and producing as well as owning this substance is illegal in most countries. By studying a case where the knowledge monopoly of the state to create and prescribe a drug is challenged, Söderberg shows that processes of innovation cannot be viewed as autonomous or separate from the state.

In their chapter in the edited volume *How Users Matter*, Dale Rose and Stuart Blume (2003) reflect on why the nation state has been ignored in most studies of user-driven innovation and research. The nation state seems to be the blind spot in these studies. The figure of the 'user' thus becomes an individual that stands outside the boundaries of the nation state as well as the established knowledge authority of universities. It seems to be assumed that if the user is free from institutional decision-making structures and authorities, the knowledge production can also flow free. Knowledge becomes synonymous with what is creative, innovative, and productive. But the 'user' is always implicated in other relations. For example, Rose and Blume (2003) discuss the users of state-sponsored vaccine programmes and how the individual user perspective is connected to the common good, such as public health, since the biosafety of the whole population is at stake.

The multifaceted relations that are implied in 'user-driven innovation', including the relation between users and the state, is hidden in the term 'user', and according to Söderberg, it is better captured with the concept of citizenship. In his study, he argues that drug users and patients form a continuum and make visible two extremes of the state: the repressive and the caring. The study was conducted with a focus on the subculture of the European psychonauts, who explore altered states of consciousness, especially through hallucinatory drugs. It is a culture and movement with global reach and with strong connections to the 1960s American hippie culture. The empirical basis for this case study was observations and interviews collected during the attendance of four psychonaut conferences in Milan, Amsterdam, Berlin, and London from 2011 to 2013.

The repressive side of the state is exemplified by the fact that the state is in charge of deciding over the continuum of legal and illegal drug users. In connection to treatment of medical diagnoses, non-medical (and thereby illegal) substance use is also established. The authority to establish a diagnosis on which legal use of drugs can be based is reserved for medical doctors, who, since they are certified experts, are seen as competent by government authorities and medical university faculties. An alternative way of describing illegal drug use is that patients have been deprived the right to establish their own diagnoses and to use drugs in a medical way (Illich 1976). This is what the psychonauts claim in order to contest the epistemic authority monopolized by state-certified experts.

The psychonaut subculture, which is the focus of Söderberg's study, challenges established epistemic authorities. Freedom for the user, however, becomes risk for the consumer. Members of the subculture try to invent strategies to compensate for the lack of state-led consumer regulation and risk reduction, but with varying results. This puts focus on the state and its caring function, even in the absence of it.

In the absence of state surveillance and consumer protection, the illegal subculture must itself provide alternatives, such as spreading information, testing, and creating user-based databases. The responsibility lies with the individual, who must stay informed about new chemical combinations, particularly when dangerous substances are circulating. Accidental deaths are not seen as being caused by the dangerous substance per se, but by the uninformed usage of the same. In this way, risks that cannot be reduced are made culturally acceptable.

The efforts of the psychonauts in managing biochemical risks have counterparts in other areas. Citizen initiatives to measure and account for air pollution (Ottinger 2010) and microplastics in the sea (Liboiron 2016) are two frequently mentioned examples, and the former can be linked to our earlier example of air pollution. Yet another example is the nuclear power accident in Fukushima. This saw a huge response from hackers, who organized workshops on how to build instruments to measure radiation and construct user-generated databases to map contamination in exposed areas (Kera 2012). The background to this was a deep mistrust of how government authorities were informing the public about the nuclear catastrophe. Many other studies are available on how government authorities have failed to control industrial emissions and have even manipulated the measurement and accounting of emissions to hide the connection between exposure and illness (Elliot 2015).

The psychonauts case study points to a blind spot within STS and user-driven innovation, and that is the relationship between the user and the state. This study can also be seen as an analogy for resistance to vaccination, which can be understood as the position that you will not give up your individual immune system to a higher power – the state, the common good, and public health. However, with lost legitimacy of the nation state, the whole idea of citizenship is at stake.

The example of the phychonaut subculture, Söderberg argues, complements previous studies of resistance to vaccination. The psychonauts' position is in much sharper contrast to the state compared to that of vaccine opponents. The psychonauts deny that the state has any right to delimit individual freedom in the name of public health, and the state, by categorizing the drugs as illegal, places the drugs outside the sphere of state regulation.

The concept of citizen presumes a relationship between the state and the individual that should be characterized by participation, responsibility, and care. If these aspects of the relationship are dismissed – so that only the repressive side of the relationship is left – then it is, according to Söderberg, wrong to speak about scientific citizenship. Thereby the psychonauts provide an example of the *sans-papir* of the scientific community.

Food advice as popular science

In a study of a controversy concerning the Swedish National Food Agency (Livsmedelsverket) and alternative dietary experts, Andreas Gunnarsson (2019) discusses food advice as popular science. The relationships between food and health are complex, controversial, and difficult to research (Gard and Wright 2005; Shapin 2007; Guthman 2013). In addition, the fact that research results are communicated to populations in popularized ways creates problems that can be discussed from the perspective of scientific citizenship.

The mainstream idea of popular science is based on a sequence of two pillars. First, research is produced, reviewed, and interpreted. Second, it is popularized and communicated to a wider audience in a simplified yet intact form. With popularized science, however, publics are left with simplified versions of science only, and do not get access to the first pillar, 'real science'. Popular science, then, not only presents science in an understandable way, but also creates a space in which researchers and experts can criticize the simplified version, since they have access to real science while most members of the public have not. A strong prerogative of interpretation is, rather, granted to scientific experts. Popular versions of science may thus have the function of strengthening the epistemic authority of science (Hilgartner 1990; Gieryn 1999).

Popular science creates a situation for citizens in which expert knowledge is difficult to criticize: the options are limited to accepting the popular science presentation as it is or appearing ignorant and scientifically illiterate. In other words, popular presentations of science may seem to make science more available and more authoritative. However, popular science understood in this way also suggests the possibility that all critique could be rejected on the basis that the science behind is more complex. However, this option is open only for scientific experts, who can protect 'real science' by the argument that popular science consists of mere simplifications.

The Swedish National Food Agency is tasked with promoting the conditions for consumers, in particular children and young people, to make conscious choices concerning healthy and safe food. This motivates the agency to, for example, promote 'an increased consumption of vegetables, roots and legumes and a bigger share of fruits, berries, nuts and seeds in the diet', since this 'would imply huge health benefits in the population' (Eneroth et al 2014: 15). Such a statement however easily becomes a target for promoters of low-carbohydrate diets, who also rely on science in their argumentation.

The low-carbohydrate movement has become rather influential in the Swedish public debate about food and health. Even though the individuals who make themselves a voice for a low carb, high fat (LCHF) diet are heterogeneous, they still form a voice that is clearly different from that of the

National Food Agency. In short, the food advice from LCHF proponents is based on the view that obesity and related unhealth is not caused by too much fat in the food, but rather by too many carbs. The way to lose weight and achieve better health is to eat less carbs and more fat. This is a recommendation that is in stark contrast to the advice given by the National Food Agency, which emphasizes a balanced diet with a limited amount of fat.

How can different actors referring to scientific knowledge come up with such different advice and conclusions? The comparison of the National Food Agency and its critics in the study by Gunnarsson makes clear how the communicating actors picture themselves, the knowledge they refer to, and the public they aim to target. The Agency, as well as its critics, clearly want to present a science *for* the people – that is, they both give expert advice in order to promote public health. Even though there is a conflict between the agency and the LCHF proponents, neither side in this debate is interested in increasing citizen influence. The debate takes place with the basic assumption that dietary research can and should provide unambiguous recommendations based on scientific knowledge.

The National Food Agency provides various tools that people can engage with, such as films, online surveys, and digital food diaries. The advice is based on knowledge about population health, but is formulated to target individuals. Responsibility shifts to individual citizens, who are expected to use their free will and the knowledge that is communicated to them to make rational choices in their own lives. Public health is thus not something that is expected to improve through bans, legislation, or state regulation, but through people's free and rational consumer choice. The critics, on the other hand, target the advice given by the agency, claiming that it is even harmful for individual health. They say it is not based on sound science, but on strong economic interests that have corrupted both research and food advice to citizens.

One type of critique against popularized science is that it is undemocratic and technocratic, but in the debate on food advice in Sweden, the critique has been the reverse: no increased citizen influence has been demanded, but rather an expertise that can deliver the certain knowledge promised by popularized versions of the debate. The practice of translating epidemiological and statistical data on public health and food habits to popular advice targeting individual citizens creates a situation where uncertainties and differences between individuals disappear. This opens up for a critique that reintroduces uncertainties and thereby undermines the logic behind popularization.

Gunnarsson's study shows the interrelatedness of a knowledge dimension and a governing dimension. The advice constructed in relation to public health problems addresses individuals rather than the population. This moves the responsibility for public health from the state to the individual, who is expected to listen to the information and make the right choices in order to contribute to both their own health and the health of the population. Public health is thus not something to be improved through legislation or other state regulations, but rather through people's voluntary and rational consumer choices (see also Petersen and Lupton 1996 and Foucault 2013).

The state as present through its non-presence

The state can be involved as an expert (through expert authorities), but it is also an important actor when it comes to organizing the boundaries between scientific experts and other actors in society (Jasanoff 2005). Mark Elam and Margareta Bertilsson (2003: 238) describe a move away from the 'enlightenment' state, which sees its task as guaranteeing the autonomy of science, to a state that encourages close cooperation between science and other actors. Even though these tendencies that Elam and Bertilsson set out more than 20 years ago are today even stronger, the state acts through various modes and guises. The enlightenment state thus still exists but in parallel with other types of state involvement.

We find a clear example of the enlightenment state in the case of the Swedish National Food Agency and its advice to the population. However, these ambitions to govern through rather detailed information and advice have, to a large extent, been replaced by other forms of governance. Often this is about citizens being involved in defining the situation based on their own individual needs. Even if the state can be an important actor(s) (alluding to the multiple levels of the state) in organizing the boundaries between science and society, the state cannot be conceived as *one* coherent actor. How the relation between experts and citizens manifests depends on the history and practices of societal institutions, shaped by legislation as well as other materialized infrastructures.

The state governs more directly – for example, by facilitating how several knowledge producers can contribute, as we saw in the earlier example of air quality – as well as indirectly – as we saw in the case of illegal drug-production. Therefore, we need to acknowledge how state involvement varies according to several dimensions. For example, one dimension concerns how the state cares for and communicates with its citizens, another dimension concerns how the state organizes knowledge production and invests in science and innovation.

These dimensions can take different directions. For example, research can take place in collaboration between universities and businesses, without citizens being actively involved in the production of knowledge. Hybrid knowledge production can take place between industry, universities, and government, and is based on the view that citizens are passive subjects of education and information campaigns. The linear knowledge model is clearly illustrated in the study on dietary advice but also, to some extent, in the example of air quality. This face of the state – the enlightenment state – has been the cornerstone of the discussion on how to organize a democratic welfare state that utilizes expert knowledge for the benefit of its citizens – that is, a science *for* the people. Our point is that this type of governing still exists but it operates in parallel to other forms of governing.

The citizen as the end point of a knowledge-based governance process becomes an object for others (experts and the state) to act on and try to shape according to the others' own framings. But this does not necessarily mean that people identify themselves with these narrow roles and become receptive to this type of control (Soneryd and Uggla 2015). As we have seen in some of the examples provided, citizens can always enrich expert knowledge by placing it in a broader context and acting as both knowledge and political subjects. Efforts to govern in particular ways will be contested when citizens do not accept the passive role assigned to them. But citizens' insight into and ability to influence knowledge production and decision-making processes will vary depending on how boundaries have been drawn between science and politics, and depending on how open decision makers are to citizens' protests and alternative understandings.

There are also many current examples of citizens who are involved in the shaping of themselves as knowledgeable and political actors. The environmental movement is an important example, being motivated by a critique of the linear knowledge model; by highlighting citizens' knowledge and experiences, movement representatives can criticize technocratic politics. The direction of this educational movement is thus the opposite of what happens in the linear knowledge model, based on the idea of public deficits. In this movement, it is citizens who act collectively to influence experts and decision makers, who dominate established knowledge production and governance. But whether the messages from social movements receive responses or not is an empirical question. In other words, it is an open question whether the involvement of citizens as subjects of knowledge and political subjects affects knowledge production and governance. The issue of climate change is still characterized by collaboration between states, and between states and business actors, even if many national government representatives talk about an inclusive climate transition. This type of collaboration can impede citizens' transparency and opportunities for influence.

As an example, in the Swedish climate policy framework, we find an overall guiding target of being a fossil-free society by the year 2045. This should be carried out by a transformation of society in a way that is 'inclusive' – making it possible for everyone to participate in the work of changing the society – and builds on legitimacy, trust, justice, and acceptance (Swedish Government 2019: 48). All citizens need to be engaged and should feel themselves engaged in the process of transforming society. In short, they should be engaged and knowledgeable climate citizens. The inclusive transformation is a political aim. However, much remains to be done to make these ideas real. Tendencies towards relying on technical fixes, such as electric vehicles or fossil-free industrial production, can counteract an inclusive climate transition that relies on citizen engagement.

Conclusions

Scientific citizenship is a concept in STS which has been given meaning by empirical studies of the ways in which scientific knowledge forms the basis for powers exercised by established governing bodies and counter-powers demonstrated through citizen contestation and mobilization around issues and citizen-led knowledge practices. As such, the concept highlights the intrinsic relation between knowing and governing. In addition, we argue that the term also emphasizes the role of the state, and should do so more explicitly.

The three examples of various forms of scientific citizenship presented in this chapter help us to locate scientific citizenship in a discussion about democracy. First, they highlight the contextual aspects and that democracy is indeed best understood as a patchwork – of regulations, of concerns about the health or safety of entire populations, of ways of facilitating or impeding the possibilities to get insight into public decision-making, and of making concerns heard and, hence, influencing public life. Second, materiality is a feature of both the efforts of established bodies to govern and of issue publics and citizen-led participation. The sheer difference in size and resources often makes the nation state, and multinational corporations, more powerful in accomplishing and distributing the material infrastructure for knowledge production compared to citizen-based activities. The examples of open and user-driven innovation and citizens mobilizing counter-expertise and building their own instruments for measurement show that such power relations are not given, and they are certainly not unidirectional and coordinated in ways that predetermine outcomes. Third, the state is present in these processes also through its non-presence. By drawing boundaries around the legal and illegal, nation states also have the power to decide what falls into the domains of its own regulations, and when the regulation of health and safety falls outside the scope of the nation state, it also falls within the scope of individual citizens or actors in the market.

The variation in how scientific citizenship is enacted can be captured by an emphasis on the relationship between experts and citizens, as well as the relations between the state, university-based research, and industry. This double focus allows us to interrogate how modes of knowledge production may be far-reaching when it comes to integration between science and policy, but this research focus still offers very few insights to concerned groups of the public (cf the two delegations constituting delegative democracy discussed in Chapter 2).

Scientific citizenship points to the fact that knowledge is always intertwined with governance, but in various ways. Citizenship needs to be explored in relation to how governing bodies imagine and respond to citizens as well as in relation to how individuals or concerned groups of the public react to governing bodies and their framings. The former can be both repressive and caring, a source of domination as well as of empowerment. These aspects of citizenship are best captured with a distinction between the historical and formal aspects of citizenship, the roles and identities citizens ascribe to as well as acquire through acts of citizenship, which may aim to discuss, question, and extend rights and obligations.

8

What Can Science and Technological Studies Say about Science and Democracy?

Introduction

Throughout this book, we have discussed a range of challenges and possibilities that arise in relation to science and expertise. These include the boundaries between science and democracy, or more generally between science and politics, and how concerned groups or the general public can be involved in technoscientific issues. It is almost impossible to find areas in life that has not been influenced by technoscientific developments. When the results of science inflict on people's lives and when political decisions and public discussions are framed by scientific expertise, fundamental and classical issues concerning democracy emerge: Who is concerned? Who governs? Who benefits? What negative impacts and uncertainties are to be found? Our focus on scientific knowledge and expertise could be assessed as a limitation in our approach to democracy, however we argue that science and technology developments inevitably shape social order and therefore also impinge on how we can imagine and practice democracy. In this final chapter, we present our view on how STS can contribute to an improved understanding of the interplay between science and democracy.

In the introduction to this book, we referred to the shadow theory of democracy, which points to the hidden assumptions and unexplored premises of democratic theory. These hidden premises can also be thought of as vulnerabilities of todays' liberal democracies concerning questions such as: Who is the public? Who governs? How can we guarantee that democracy leads to equal outcomes? (Gustavsson 2018). These are termed 'vulnerabilities' because they are often assumed to be connected to clear answers. But they seldom are, and they can, therefore, lead to mistrust in the democratic system altogether. They can also pave the way for charismatic leaders that give clear answers to these questions, even though these types of answer do not really exist. The public is not easily demarcated, since those who are concerned by decisions cannot always be identified in the population within a nation state. Moreover, those who govern are far from limited to publicly elected politicians, but include a range of individual and organizational decision makers, and they may have to deal with lock-ins that are difficult to change and hence claim control over. In addition, democracy is about equal rights and opportunities, and if formal democracies fail to deliver fair and just outcomes, this will also lead to mistrust of the current system and pave the way for populist movements that promise better results.

Thus, if the vulnerabilities are ignored, they may lead the way towards populism – that is, non-pluralism (Gustavsson 2018). Therefore, we need to pursue a continuous and deepened dialogue about the vulnerabilities of democracy. We argue that STS has strong contributions to make in such dialogues, concerning both who governs and how the public is demarcated.

One conclusion that STS researchers draw from their studies about developments in science and technology is that 'it could always be seen otherwise' (Woolgar 2014: 31). The knowledge produced in a scientific practice will be different if the context changes, which can lead to new combinations of problems, questions, actors, and material objects. But if STS researchers do not want to stop at *describing* how science works in practice, on what basis can they claim that it is possible to distinguish between and assess various knowledge claims? On what basis is it possible to claim that some knowledge is more correct than others? This is an additional issue that we take a clearer position on in this final chapter.

In the following, we address four themes that we have already discussed in this book, summarizing and taking a stance in relation to each one. These themes are important ongoing discussions within STS, which we try to make some conclusions about.

The first theme concerns *representation* and *participation*. We argue that the issue of participation comes to the fore when an improved – better – representation is under discussion. But does participation always improve representation, and how can we assess when it does? Representation and participation connect to the question of how to demarcate publics as well as the question of how to distinguish between different knowledge claims. With a relational view of representation, however, these questions become intertwined. If representation concerns both facts and publics at the same time, we can attend to the consequences of the two elites (elected politicians and scientific experts): an ignorant mass and the risk of overflows. The challenge then is to assess what improved participation means for facts and publics. This is about assessing the *quality* of representation and participation.

In our second theme, we discuss *hybrids* and *modes of governance*. Callon and his co-authors (2009) distinguish between hybrid forums that are cultivated and those that are not. Hybrid forums emerge spontaneously (when overflows occur) but can then be taken care of in different ways. They are often met with resistance, ignored, or denied, but they can also be reflected on and utilized for changes and improvements. This theme is related to but does not totally overlap the question of participation. It is important to see that hybrid forums produce both alternative knowledge and alternative modes of governing – they are both counter-expertise and counter-democracy (Rosanvallon 2010). We argue against a too naive stance for hybrid forums, since they, in their uncultivated and spontaneous forms, might just add layers of power and mechanisms of exclusion. We then need to assess the *quality* of existing hybrids.

Third, there is a debate, even if not very prominent so far, about the relationships between STS, democratic theory, and political philosophy. In this debate, it has been argued that STS should become better at relating to political philosophy and even to general social theory. Several STS scholars (for example, Marres 2007; Wynne 2007; Brown 2009; Durant 2011) have ideas on how this can be done, but they propose different ways. We partly agree with what is formulated in these proposals, but we also want to claim that STS has a unique contribution to make to the analysis of the interplay between science and democracy precisely because of its many detailed empirical studies.

The fourth theme is about the normativity of STS. Is there a 'built-in' normativity in STS? STS is based on a view of knowledge that recognizes that representation is not something simple, and this view has consequences for STS scholars' emphasis of the importance of participation. We summarize and make conclusions, based on this normativity, about what we believe STS can contribute to in a discussion about science and democracy.

Representation and participation

How can we distinguish between good and bad knowledge, and good and bad governance? Assumptions that we have related to throughout the book are that scientific knowledge is the best knowledge and democracy is the best mode of governance. But this only solves questions at the level of labelling. Whether a representation is good and fulfils its specific purposes, or if there are deficits and inaccuracies, must always be assessed in practice and in relation to specific issues. This is where STS research makes a clear connection between representation and participation. When participation is exclusive and limited to a few experts, representation becomes incomplete. However, this is the crucial point with representations – they are meant to simplify. Science is characterized by simplification and reinforcement of certain features, what Latour (1999: 71–72) describes as a series of transformations involving both reduction and amplification: to reduce means to amplify. To select some aspects and make them more important than others is one of the central characteristics of science. People who are affected by the results of the simplified constructions of experts – for example, due to health or environmental effects that the experts have not predicted or fully understood – can draw attention to consequences they experience and make those known to responsible experts and decision makers. And this is done all the time by colleagues in the scientific community, by affected local populations, but also by people who mobilize and engage in specific issues – through social movements – and who pay attention to limited and incorrect representations of human bodies, distribution of resources, or environmental effects.

Another way of putting this issue is that good representations are representations that are perceived as legitimate by the groups and populations affected by the consequences of actions stipulated by these representations. But is good representation always the same as legitimacy? Collins and Evans (2007) argue that knowledge must also be credible. This means that representations cannot only be seen in relation to the legitimacy of a wider population. According to Collins and Evans, the representation must also be judged to be good by the small group that (usually) knows the most about the matter in question. Here, however, we encounter problems if, at the same time, we accept that those people (often scientific experts but not always) are often not the best at discovering their own blind spots.

Through close empirical studies, STS has drawn attention to the limitations of the exclusive groups of experts who have much influence over plans and decisions that affect many, and therefore the STS position has often been in favour of extended participation, since an extended group of participants presumably can attend to aspects that are missing and should be considered. Through broadened participation, issues can be understood and managed in different, and sometimes better, ways. However, if participation is extended to include a number of concerned or affected groups, paralysis can occur, and if relevant expertise is excluded, this will also lead to overflows. The fact that certain issues are expert dependent means that experts *are* needed. There is always a necessary tension between the simplification of issues (specialization) and extended participation. We suggest that a good representation is a composition – that is, a combination of ideas, people, and things – that is so stable that no one demands a revision of its framing. Then, it does not attract attention because it is too narrowly defined to address the problems it is intended to solve, and it does not misjudge essential aspects so that overflows trigger new questions and problems. However, a representation that is both legitimate for a wider public and credible among the smaller groups of experts may still be deficient. This is because not everyone can make themselves heard and because problems and overflows can show up at a later stage.

Extended participation can potentially mean that shortcomings of expert framings and overflows are noticed, and the representation can thus be improved if the shortcomings are remedied or overflows are recognized and perceived as such, and hence lead to reframing the entire problem complex. But this is dependent on the participation being extended to committed groups that do just that – pay attention to shortcomings that experts have not seen – and on decision makers being prepared to react to the shortcomings and overflows that uninvited groups are paying attention to. STS research on extended participation shows that the question of how concerned groups of the public are delimited is far from obvious. Above all, the public is not a stable and fixed entity, but emerges in relation to the issues that engage the public.

This *issue-oriented* and *problem-centred* view of the public means that STS scholars draw attention to the fact that there is a multiple and diverse set of affected groups. STS researchers are thus often critical of authorities and private organizations that appoint themselves as interpreters of public views and attitudes by claiming that they have consulted the public. The issue of good representation can be summarized as an ability to both pay attention to and manage overflows. This means that issues that were previously perceived as separate, based on their technical and social dimensions, are now perceived as hybrid issues. But are hybrids always the solution? We suggested earlier that representation can be assessed by looking at how it is composed: a stable composition means that nobody – neither relevant experts nor concerned publics – is questioning the framing of it. Paradoxically, however, in a new age of open innovation, described in Chapter 4, the compositions may have changed so that it is even harder to de-compose them. This can be the case if these are primarily market oriented and not run according to principles of openness in all respects. Overflows of information and a more varied media landscape might also make it more difficult for concerned publics to know what is going on, even in their own vicinities. Compositions that are more interconnected – through, for example, a financialization of the economy and digitalization – might also be harder to de-compose.

We can draw the following conclusion in relation to STS approaches to representation and participation: the many empirically detailed STS studies together with an open-ended question about how things are combined, questioned, and changeable are important contributions to our understanding of how power and counter-power emerge, how concerned publics mobilize around issues, and how governing bodies may respond to these concerned publics. These are studies of how both science and democracy work in practice: how knowledge authority is gained or refuted, who gets influence and how, and who is concerned and perceived or dismissed as legitimate participants. Social theory in general, and democratic theory in particular, can gain from more studies of how things work in practice, and hence from STS insights.

STS, on the other hand, often focuses on detailed case studies and less commonly makes analyses of overarching changes that go beyond the particular case (although important exceptions exist). We believe that STS could gain by zooming out from particular cases to look at the geopolitical, regulatory, and historical contexts. Since this is about *context* – rather than theory that is secluded from the empirical – it might be possible to say something more general about the conditions for public scrutiny of established compositions, in order to assess the quality of existing hybrids from a democratic point of view.

The danger of celebrating hybrids

Ideas about separation often lead to misguided political measures, since practices are characterized by hybridization. Callon and his co-authors (2009) believe that because of this, the solution can be found in hybrid forums. But an unsorted praise of hybrids also involves a risk, at least if the idea of hybrids is based on the notion of a flat world, without hierarchies and already established power relations (cf Latour 1996). However, Callon and his colleagues are aware of this:

> To leave hybrid forums to develop without any rules of the game for organizing the debate would leave the field free to the logic of relations of force, it would allow the reproduction, without discussion, of the exclusion of the weakest, precisely all those who seek to make their voices heard and be listened to. (2009: 154)

To leave the hybrids to themselves means that we are leaving the field free for power relations to take shape. There is always the risk that existing inequalities are reproduced and that new ones are created. Brown (2009) places his hope in democratic institutions that can even out inequalities. It is hard to refute that it is better to live in societies that have institutions that guard citizens' right to information about plans and decisions that concern them, and freedoms of expression and of assembly and association, than to live in societies that do not grant these rights. Equally, we might think it is better to live with institutions that sometimes ignore and sometimes respond to citizens' expressions of protest than with institutions that try to silence these expressions through force or respond with violence. At the same time, there is no other way to assess whether these institutions are actually living up to this in one way or another than to study them in practice, and this is a question of how we define the boundaries of societies and concerned publics. In the study of practices, we will find that there are ongoing processes of both separation and hybridization. Analyses that aspire to avoid a flat view of the world and power relations thus need to study the consequences of both separation and hybridization, in terms of inclusion and exclusion.

Practices also need to be attended to when it comes to democracy. Democracy can never be taken for granted and must always be done again and again; in other words, democracy is practiced and must always be practiced. STS that has a focus on practices would perhaps benefit more from discussing democratization as a verb (something that is done) rather than democracy as a noun (something that is filled with a specific content). The overflows that publics pay attention to require a response and, thus, demand some kind of change by governing bodies. These governing bodies must react to the attention to overflows, but as soon as this is done, new framings will be created and, hence, new possible overflows. No framing is completely tight, and never in the long run. Democracy is, thus, always temporary and arises only at the moment when the unrepresented resist and when this is noticed by those in power. The conceptual pair of framing and overflows, from Callon, are reminiscent of the ideas of Jacques Rancière.

Rancière's (2009) concept of democracy emphasizes that it is in the confrontation between those who do not count and the established order that democracy temporarily arises. But difficult as it is to assess the quality of institutions and organizations without studying them in practice, it is just as difficult to be an omnipotent interpreter of the moment when legitimate demands are met in the form of a (correct) revision of current modes of governance. STS researchers must, therefore, always be reflexive about their own analyses and expose them to the possibility that their research objects protest and resist what is said about them.

One way of assessing the quality of hybrids is to focus on what participation is about. Callon et al (2009) propose the notion of 'technical democracy' as an alternative to delegative democracy. When overflows occur, for whatever reason, the issue in question needs to be reframed, or at least the overflows need to be taken care of, and seldom can this be done by just deflecting them without modifying the issue at stake. Reframing the issue, however, also involves participation. As STS researchers, we should be interested in what this new and reframed participation involves. Technical democracy is a proper name for dealing with overflows in relation to technical issues that have been questioned, often by outsiders, due to their (unintended) consequences. When these outsiders, or concerned laypeople, question representation of both the issue and the participation, the quality of possible new emerging issues and participants should be assessed from the angle of technical problematization – that is, the technical object as the problem in need of reassessment and modification. If the technical object is not opened up for (new) participants to discuss and negotiate, the quality of participation is low. Extended participation should be seen as a tool that feeds technical problematization. This is the only way to achieve technical democracy (Barthe et al 2022).

When local inhabitants question the siting of a final repository for nuclear waste, overflows occur, which could be taken care of in different ways. If the focus in this modifying work is on the protests as such, the worries among the inhabitants, the procedures for decision-making in relation to a proposed application – that is, on the *social issues* – the quality of the participation as well as of the new emerging hybrids is lower compared to a situation in which *the technical object* itself – the waste, the methods for storing the waste, and the technical calculations about the storage – is reframed as a result of new participants being part of the negotiations. An old terminology for this difference in what participation is about is *token participation versus real influence*.

One important conclusion is that STS can contribute to an in-depth discussion about who or what governs by looking at both separation and hybridization – for instance, between the social and the technical – and their effects in real situations. Different constellations and mixes of expert rule and political rule are often the focus of STS analyses. Rather than seeing these hybrids as either good or bad in themselves, their *effects* should be attended to. In relation to specific compositions, we should ask the following questions: Who or what is represented? What are the benefits of this composition in terms of consequences (better environment, increased standard of living, increased recognition for the unrepresented, and so on)? What are the conditions for recognizing and attending to overflows?

Studies of situated practices, theories of democracy, and political philosophy

STS studies are not void of theory, but they involve theory in more empirical and practical ways than conventional studies that distinguish between theory and the empirical. But what are the consequences for STS and this way of relating to theory in analyses of the interplay between science and democracy? If there is a kind of 'hidden' concept of democracy in STS research, it might be important to make this clear. For example, Brown (2009) has discussed how STS researchers prefer to discuss participation rather than discussing democratic institutions. Does this mean that STS researchers become hidden advocates for more direct democratic or deliberative forms of democracy? One topic that is discussed among STS researchers is precisely this possible tension between empirical studies of practices versus democracy theory and political philosophy.

Darrin Durant (2011) mentions Brown as one of few who try to relate STS to political philosophy. Durant identifies different models of democracy behind STS researchers' arguments for participation and attempts to distinguish between different forms of expertise. Based on this analysis, Durant concludes that Jasanoff's and Wynne's approaches fit relatively well into the Habermasian tradition, because of their relational ontologies, while Collins and Evans' approach is closer to the tradition to which John Rawls belongs, due to the more deductive approach – that is, relying on categories that are established even before looking at the practices and how relations are enacted. We have already said that an understanding of democracy as something temporary could suit practical STS research. Temporarily, uncultivated hybrid forums can shake things around and get some kind of response. However, the normative basis – for example, an injustice that motivates change – also needs to be made explicit, otherwise changes could easily also be the result of lobbyism or other kinds of power relations.

Models of democracy, or the theory-building around the institutions of democracy, hardly constitute a strength within STS, and perhaps this should not be the case either, since others are better at this. The strength of STS research is, rather, studies of how epistemic and political authority are established, questioned, and possibly transformed. And in our view, this is about studying both representation and participation as relational concepts that are filled with meaning in empirical cases. We believe that this is what STS does. The specific contribution we find in STS research in terms of analyses of the interplay between science and democracy is precisely the contextual and empirical one.

Still, we also agree with Wynne (2016) that STS needs to relate to political philosophy in order to develop a more relational approach and to better understand participatory processes. It is by relating to political philosophy that STS has developed a relational understanding of the public. It is when people pay attention to problems which expert authorities neglect that they engage and emerge as an 'affected' public (Marres 2007). We have suggested that a good composition is a composition that is so stable that no one requires the framing to be revised. However, to assess this, the compositions must be constantly subjected to testing and, if necessary, revised. Concerning delegation to expert groups, it is important to be aware that these are in themselves neither democratic nor non-democratic, but relational. How are they perceived? Are there those who want to make their voices heard, who want to point to failures and deficits, and who thereby confront and question an established delegation? An important question for everyone who studies democracy as a practice is also how sustainable our current configurations of democracy are. If imaginaries of democracy are linked to the nation state and its specific institutions, how do we then manage to see ourselves as intertwined with humans and non-humans on the other side of the globe? Representations can potentially be constantly expanded. The number of unrepresented is inexhaustible. This is becoming apparent to more and more people in the ongoing debate on climate change.

We have suggested that an STS approach to science and democracy is relational. Science and democracy are both patchworks of actors, practices, institutions, and materialities as well as ideas and knowledge. We believe that the relation of STS to theory is both a strength and a weakness. It is a strength when the empirical and the theoretical are placed on the same level – seen as equally valuable sources to think with – and it is a weakness when theory becomes reified as 'general theory'. Democratic theory has perhaps been seen too much and for too long as the general theory that STS cannot sit in companionship with. However, democratic theory is also just a patchwork of ideas, values, and analytical distinctions that are just as good to think with as other resources, and perhaps this is especially the case when democracy is the topic we wish to say something about.

STS and normativity

Perhaps STS researchers have been more actors than analysts, at least when they get involved in democracy issues. STS researchers have pointed out shortcomings, revealed power relations, and given a voice to those who have not yet been heard. The god trick – that scientists are omnipotent interpreters of the truth – no longer works (Haraway 1996). Today, it is not considered correct to claim that knowledge is valid regardless of context and relevance. This requires that the assumptions behind the knowledge claims, as well as the contexts in which the claims are considered to apply, are made clear. But, today, we find countless contexts where this is not done. Technological innovations that are competitive on a global market are often emphasized as the only way forward, and participation is extended to more people without the ambition of making representations available for criticism and revisions.

We see many examples of efforts to make a sharp division between the two elites – experts and political leaders – and the ignorant masses. The real problem behind post-truth and alternative facts is continued separation between science and democracy. When power is not responsive to the resistance of the unrepresented, this is a problem in relation to knowledge as well as democracy. In an advanced knowledge society, with intensified transnational exchanges of goods, ideas, and innovations, the conditions for resisting are changing, and there is uncertainty as to what or who resistance can and should be directed at. This means that science is not only a resource for dominance, but also a resource for resistance.

The view of knowledge that STS stands for is clearly normative. STS research focuses on analysing power as well as resistance, creativity, and unpredictability. Being open to both hybridization and separation requires studies of practices. This means that STS research needs to study basic co-production (or constitutive co-production) that focuses on new creations and hybrids that rearrange the world as well as controversies and power relations within already stabilized power relations – that is, interactionist co-production. Within these manifested power relations, the attention to overflows can lead to important changes (and improvements) even if they do not overturn existing and established power relations. STS has, since its beginnings, been a co-actor in this awareness of shortcomings and overflows, and at the same time it has contributed to analyses that help us to better understand science in society. It is in this perspective of science – and how it relates to action, governance, and democracy – that we find a normativity. It is in this way that the interplay between science and democracy should be studied.

Normativity, however, also means that we need to take a stance in relation to what values we want to protect, or achieve through the best knowledge (science) and the best governance (democracy). The 'rule of the people' is not a value in itself to be achieved, but rather a means to achieve some degree of individual freedom, justice, social equality, and non-violence.

Actors are important to STS, and thereby it is natural for STS scholars to focus on and try to improve participation. We can conclude that the importance of participation is an intrinsic, theoretically based normativity in STS. In this book, and not least in this last chapter, we have argued that by focusing on participation, STS research can open up a critical and empirically based understanding of representation of issues and actors (including publics), and thereby of participation itself. Concepts such as co-production, hybridity, hybrid forums, and technical democracy provide important support to such an ambition.

Conclusions

Should STS come out of the shadows and say anything substantial about democracy, good governance, and how to assess the quality of democratic institutions? We argue that STS should remain in the shadows of democratic theory (see Chapter 1) but that it could be more reflexive about what this could mean for the improvement of democracy. The key contribution of STS is empirical, however, and this empirical approach is also how STS contributes to theory and theorizing.

STS adopts a view on knowledge which implies that representation is not a simple thing, and this has consequences for how the role of public participation in technoscientific issues is assessed. This is a *relational* specification of good and bad: whenever there is someone challenging a representation which is 'a bad' for someone, the struggles over good and bad become apparent; hence, there is a need to attend to overflows. Science is not only a resource for domination; it is also a resource for resistance. More engagement with democratic theory could, however, help to make the normative grounds for attending to overflows more explicit.

How then can STS make its normative standpoints more explicit? By developing an approach to science as well as democracy as the continuous configuration of processes, STS can contribute with its specific perspective on science and democracy – as emerging, relational, and unfinished, and always changing.

References

Ackerman, Bruce & Fishkin, James S. (2004). *Deliberation Day*. New Haven, CT: Yale University Press.

Agrawala, Shardul (1998). Context and Early Origins of the Intergovernmental Panel on Climate Change. *Climatic Change* 39(4): 605–620.

Árnason, Vilhjálmur (2012). Scientific Citizenship in a Democratic Society. *Public Understanding of Science* 22(8): 927–940.

Arnstein, Sherry R. (1969). A Ladder of Citizen Participation. *Journal of the American Institute of Planners* 35(4): 216–224.

Asdal, Kristin and Hobæk, Bård (2016). Assembling the Whale: Parliaments in the Politics of Nature. *Science as Culture* 25(1): 96–116.

Asdal, Kristin, Brenna, Brita & Moser, Ingunn (eds) (2007). *Technoscience: The Politics of Interventions*. Oslo: Unipub.

Barnes, Barry & Edge, David (eds) (1982). *Science in Context: Readings in the Sociology of Science*. Milton Keynes: Open University Press.

Barthe, Yannick, Elam, Mark & Sundqvist, Göran (2020). Technological Fix or Divisible Object of Collective Concern? Histories of Conflict over the Geological Disposal of Nuclear Waste in Sweden and France. *Science as Culture* 29(2): 196–218.

Barthe, Yannick, Meyer, Morgan & Sundqvist, Göran (2022). Technical Problematisation: A Democratic Way to Deal with Contested Projects? *Science, Technology and Society* 27(1): 7–22.

Beck, Silke (2011). Moving Beyond the Linear Model of Expertise? IPCC and the Test of Adaptation. *Regional Environmental Change* 11(2): 297–306.

Beck, Silke (2012). Between Tribalism and Trust: The IPCC Under the 'Public Microscope'. *Nature and Culture* 7(2): 151–173.

Beck, Ulrich (1992). *Risk Society: Towards a New Modernity*. London: Sage.

Bijker, Wiebe E. (1995). *On Bikes, Bakelites and Bulbs: Towards a Theory of Socio-Technical Change*. Cambridge, MA: MIT Press.

Blok, Anders (2022). What is Democracy According to STS? *Science as Culture* 32(1): 156–160.

Bloor, David (1991 [1976]). *Knowledge and Social Imagery* (2nd edn). Chicago, IL: University of Chicago Press.

Blue, Gwendolyn and Medlock, Jennifer (2014). Public Engagement with Climate Change as Scientific Citizenship: A Case Study of World Wide Views on Global Warming. *Science as Culture* 23(4): 560–579.

Blühdorn, Ingolfur and Deflorian, Michael (2019). The Collaborative Management of Sustained Unsustainability: On the Performance of Participatory Forms of Environmental Governance. *Sustainability* 1(4): 1189.

Bolin, Bert (2007). *A History of the Science and Politics of Climate Change: The Role of the Intergovernmental Panel on Climate Change.* Cambridge: Cambridge University Press.

Bourblanc, Magalie, Crabbé, Ann, Liefferink, Duncan & Wiering, Mark (2013). The Marathon of the Hare and the Tortoise: Implementing the EU Water Framework Directive. *Journal of Environmental Planning and Management* 56(10): 1449–1467.

Brown, Mark (2009). *Science in Democracy: Expertise, Institutions, and Representation.* Cambridge, MA: MIT Press.

Bucchi, Massimiano (2004). *Science in Society: An Introduction to Social Studies of Science.* London: Routledge.

Callon, Michel (1998). An Essay on Framing and Overflowing: Economic Externalities Revisited by Sociology. In: Callon, Michel (ed) *The Laws of the Markets.* Oxford: Blackwell Publishers, pp 244–269.

Callon, Michel (2009). Foreword. In: Hecht, Gabrielle, *The Radiance of France: Nuclear Power and National Identity after World War II.* Cambridge, MA: MIT Press, pp xi–xxiii.

Callon, Michel & Law, John (1982). On Interests and their Transformation: Enrolment and Counter-Enrolment. *Social Studies of Science* 12(4): 615–625.

Callon, Michel, Lascoumes, Pierre & Barthe, Yannick (2009). *Acting in an Uncertain World*: *An Essay on Technical Democracy.* Cambridge, MA: MIT Press.

Carayannis, Elias G. & Campbell, David F.G. (2012). *Mode 3 Knowledge Production in Quadruple Helix Innovation Systems.* SpringerBriefs in Business, Vol 7. New York: Springer.

Carleheden, Mikael (2009). Är demokrati möjligt i komplexa samhällen? *Fronesis* 29–30: 8–24.

Carson, Rachel (1962). *Silent Spring.* Mifflin: Houghton.

Casula Vifell, Åsa & Soneryd, Linda (2012). Organizing Matters: How 'the Social Dimension' Gets Lost in Sustainability Projects. *Sustainable Development* 20(1): 18–27.

Chilvers, Jason & Kearnes, Matthew (eds) (2016). *Remaking Participation: Science, Environment and Emergent Publics.* London: Routledge.

Collingridge, David & Reeve, Colin (1986). *Science Speaks to Power: The Role of Experts in Policy Making.* London: Frances Pinter.

Collins, H.M. (1981). What is TRASP? The Radical Programme as a Methodological Imperative. *Philosophy of the Social Sciences* 11(2): 215–224.

Collins, H.M. (1992 [1985]). *Changing Order: Replication and Induction in Scientific Work* (2nd edn). Chicago, IL: University of Chicago Press.

Collins, Harry (2014). *Are We All Scientific Experts Now?* Cambridge: Polity Press.

Collins, H.M. & Evans, Robert (2002). The Third Wave of Science Studies: Studies of Expertise and Experience. *Social Studies of Science* 32(2): 235–296.

Collins, Harry & Evans, Robert (2007). *Rethinking Expertise*. Chicago, IL: University of Chicago Press.

Collins Harry, Weinel, Martin & Evans, Robert (2010). The Politics and Policy of the Third Wave: New Technologies and Society. *Critical Policy Studies* 4(2): 185–201.

Collins, Harry, Evans, Robert & Weinel, Martin (2017). STS as Science or Politics? *Social Studies of Science* 47(4): 580–586.

Collins, Harry, Evans, Robert, Durant, Darrin & Weinel, Martin (2020). *Experts and the Will of the People: Society, Populism and Science*. Cham: Palgrave Macmillan.

Commission of the European Communities (2001). *European Governance: A White Paper*. COM(2001) 428.

Corner, Adam & Groves, Christopher (2014). Breaking the Climate Change Communication Deadlock. *Nature Climate Change* 4(9): 743–745.

Dahl, Emmy (2014). *Om miljöproblemen hänger på mig. Individer förhandlar sitt ansvar för miljön*. Gothenburg: Makadam förlag.

Dahl, Robert (2008). *Democracy and its Critics*. New Haven, CT: Yale University Press.

Dányi, Endre (2018). The Things of the Parliament: An ANT-Inspired Reading of Representative Democracy. In: Brichzin, Jenni, Krichewsky, Damien, Ringel, Leopold & Schank, Jan (eds) *Soziologie der Parlamente: Neue Wege der politischen Institutionenforschung*. Wiesbaden: Springer VS, pp 267–285.

Davies, Sarah R. & Horst, Maja (2016). *Science Communication: Culture, Identity and Citizenship*. London: Palgrave Macmillan.

Delfanti, Alessandro (2013). *Biohackers: The Politics of Open Science*. London: Pluto Press.

Demortain, David (2020). *The Science of Bureaucracy: Risk Decision-Making and the US Environmental Protection Agency*. Cambridge, MA: MIT Press.

De Pryck, Kari & Hulme, Mike (eds) (2022). *A Critical Assessment of the Intergovernmental Panel on Climate Change*. Cambridge: Cambridge University Press.

Dilling, Lisa & Lemos, Carmen Maria (2011). Creating Usable Science: Opportunities and Constraints for Climate Knowledge Use and their Implications for Science Policy. *Global Environmental Change* 21(2): 680–689.

Durant, Darrin (2011). Models of Democracy in Social Studies of Science. *Social Studies of Science* 41(5): 691–714.

Durant, Darrin (2016). The Undead Linear Model of Expertise. In: Heazle, Michael & Kane, John (eds) *Policy Legitimacy, Science and Political Authority: Knowledge and Action in Liberal Democracies*. Abingdon: Routledge, pp 17–38.

Durkheim, Émile (2014 [1895]). *The Rules of Sociological Method: And Selected Texts on Sociology and Method*. New York: Free Press.

Eckersley, Robyn (2020). Ecological Democracy and the Rise and Decline of Liberal Democracy: Looking Back, Looking Forward. *Environmental Politics* 29(2): 214–234.

Edwards, Paul N. (2012). Entangled Histories: Climate Science and Nuclear Weapon Research. *Bulletin of the Atomic Scientists* 68(4): 28–40.

Eisenack, Klaus, Moser, Susanne C., Hoffman, Esther, Klein, Richard J.T., Oberlack, C., Pechan, Anna, Rotter, Maja & Termeer, Catrien J.A.M. (2014). Explaining and Overcoming Barriers to Climate Change Adaptation. *Nature Climate Change* 4: 867–872.

Elam, Mark & Bertilsson, Margareta (2003). Consuming, Engaging and Confronting Science: The Emerging Dimensions of Scientific Citizenship. *European Journal of Social Theory* 6(2): 233–251.

Elliot, Kevin (2015). Selective Ignorance in Environmental Research. In: Gross, Matthias & McGoey, Linsey (eds) *Routledge International Handbook of Ignorance Studies*. London: Routledge, pp 165–173.

Eneroth, Hanna, Björck, Lena & Brugård Konde, Åsa (2014). *Bra livsmedelsval baserat på nordiska näringsrekommendationer 2012*. Livsmedelsverkets rapportserie 19/2014. National Food Agency, Sweden.

Engelhardt, H. Tristram, Jr & Caplan, Arthur L. (eds) (1987). *Scientific Controversies: Case Studies in the Resolution and Closure of Disputes in Science and Technology*. Cambridge: Cambridge University Press.

Eurobarometer (2021). *European Citizens' Knowledge and Attitudes Towards Science and Technology*. Available at: https://europa.eu/eurobarometer/surv eys/detail/2237

European Commission (2003). *Common Implementation Strategy for the Water Framework Directive (2000/60/EC). Guidance document no. 8. Public participation in relation to the Water Framework Directive*. Office for Official Publications of the European Communities.

European Commission (2019). *Open Science*. Fact sheet. Available at: https://ec.europa.eu/info/sites/default/files/research_and_innovat ion/knowledge_publications_tools_and_data/documents/ec_rtd_factsh eet-open-science_2019.pdf

European Commission (2020). *Strategic Plan 2020–2024 – Research and Innovation.* Directorate-General for Research and Innovation. Available at: https://research-and-innovation.ec.europa.eu/strategy/strategy-2020-2024_en#strategy-document

EU (European Union) (2000). *Water Framework Directive 2000/60/EEC of the European Parliament and of the Council of 23 October 2000 establishing a framework for Community action in the field of water policy.*

Eyal, Gil (2013). For a Sociology of Expertise: The Social Origins of the Autism Epidemic. *American Journal of Sociology* 118(4): 863–907.

Ezrahi, Yaron (1990). *The Descent of Icarus: Science and the Transformation of Contemporary Democracy.* Cambridge, MA: Harvard University Press.

Ezrahi, Yaron (2012). *Imagined Democracies: Necessary Political Fictions.* New York: Cambridge University Press.

Fagerberg, Jan, Fosaas, Morten & Sapprasert, Koson (2012). Innovation: Exploring the Knowledge Base. *Research Policy* 41(7): 1132–1153.

Felt, Ulrike (2015). Keeping Technologies Out: Sociotechnical Imaginaries and the Formation of Austria's Technopolitical Identity. In: Jasanoff, Sheila & Kim, Sang-Hyung (eds) *Dreamscapes of Modernity: Sociotechnical Imaginaries and the Fabrication of Power.* Chicago, IL: Chicago University Press, pp 103–125.

Felt, Ulrike & Wynne, Brian (2007). *Taking European Knowledge Society Seriously: Report of the Expert Group on Science and Governance to the Science, Economy, and Society Directorate.* Brussels: European Commission, Directorate-General for Research and Innovation.

Felt, Ulrike & Fochler, Maximilian (2010). Machineries for Making Publics: Inscribing and De-Scribing Publics in Public Engagement. *Minerva* 48(3): 219–238.

Felt, Ulrike & Irwin, Alan (eds) (2023). *Encyclopedia of Science and Technology Studies.* Cheltenham: Edward Elgar.

Felt, Ulrike, Barben, Daniel, Irwin, Alan, Joly, Pierre-Benoît, Rip, Arie, Stirling, Andy & Stöckelová, Tereza (2013). *Science in Society: Caring for Our Future in Turbulent Times.* Science Policy Briefing 50. Strasbourg: European Science Foundation.

Felt, Ulrike, Fouché, Rayvon, Miller, Clark A. & Smith-Doerr, Laurel (eds) (2017). *The Handbook of Science and Technology Studies.* Cambridge, MA: MIT Press.

Fischer, Frank (2009). *Democracy and Expertise: Reorienting Policy Enquiry.* Oxford: Oxford University Press.

Fishkin, James S (2011). *When the People Speak: Deliberative Democracy and Public Consultation.* Oxford: Oxford University Press.

Foucault, Michel (2013). *Biopolitikens födelse: Collége de France 1978–1979.* Stockholm: Tankekraft.

Frank, David John & Meyer, John W. (2020). *The University and the Global Knowledge Society*. Princeton, NJ: Princeton University Press.

Fraser, Nancy & Honneth, Axel (2003). *Redistribution or Recognition? A Political-Philosophical Exchange*. London: Verso.

Funtowicz, Silvio O. & Ravetz, Jerome R. (1993). Science for the Post-Normal Age. *Futures* 25(7): 739–755.

Fustel De Coulanges, Numa Denis (1980). *The Ancient City: A Study on the Religion, Laws, and Institutions of Greece and Rome*. Baltimore, MD: Johns Hopkins University Press.

Gabrys, Jennifer (2016). *Program Earth: Environmental Sensing Technology and the Making of a Computational Planet*. Minneapolis, MN: University of Minnesota Press.

Gard, Michael & Wright, Jan (2005). *The Obesity Epidemic: Science, Morality and Ideology*. London: Routledge.

Garsten, Christina, Rothstein, Bo & Svallfors, Stefan (2015). *Makt utan mandat. De policyprofessionella i svensk politik*. Stockholm: Dialogos.

Gaskell, George, Allum, Nick & Stares, Sally (2003). *Europeans and Biotechnology in 2002*. Eurobarometer 58.0. A report to the EC Directorate General for Research from the project 'Life Sciences in European Society' QLG7-CT-1999-00286. London: London School of Economics and Political Science.

Gaskell, George, Allansdottir, Agnes, Allum, Nick, Castro, Paula, Esmer, Yilmaz, Fischler, Claude, Jackson, Jonathan, Kronberger, Nicole, Hampel, Jurgen, Mejlgaard, Niels, Quintanilha, Alex, Rammer, Andu, Revuelta, Gemma, Stares, Sally, Torgersen, Helge & Wager, Wolfgang (2011). The 2010 Eurobarometer on the life sciences. *Nature Biotechnology* 29(2): 113–114.

Gieryn, Thomas F. (1983). Boundary-Work and the Demarcation of Science from Non-Science: Strains and Interests in Professional Ideologies of Scientists. *American Sociological Review* 48(6): 781–795.

Gieryn, Thomas F. (1995). Boundaries of Science. In: Jasanoff, Sheila, Markle, Gerald, Peterson, James & Pinch, Trevor (eds) *Handbook of Science and Technology Studies*. Thousand Oaks, CA: Sage, pp 393–443.

Gieryn, Thomas F. (1999). *Cultural Boundaries of Science: Credibility on the Line*. Chicago, IL: University of Chicago Press.

Gould, Kenneth A. (2015). Slowing the Nanotechnology Treadmill: Impact Science versus Production Science for Sustainable Technological Development. *Environmental Sociology* 1(3): 143–151.

Goven, Joanna (2006). Processes of Inclusion, Cultures of Calculation, Structures of Power: Scientific Citizenship and the Royal Commission on Genetic Modification. *Science, Technology & Human Values* 31(5): 565–598.

Group of Chief Scientific Advisors (2019). *Scientific Advice to European Policy in a Complex World*. Scientific Opinion No 7. European Commission, Directorate-General for Research and Innovation. Luxembourg: Publications Office of the European Union.

Grundmann, Rainer & Rödder, Simone (2019). Sociological Perspectives on Earth System Modeling. *Journal of Advances in Modeling Earth Systems* 11: 3878–3892.

Gunnarsson, Andreas (2019). Kostråd som populärvetenskap. In: Soneryd, Linda & Sundqvist, Göran (eds) *Vetenskapligt medborgarskap*. Lund: Studentlitteratur, pp 45–65.

Gustavsson, Sverker (2018). Skuggteorin tydliggör problemet. In: Gustavsson, Sverker, Jonsson, Claes-Mikael & Lindberg, Ingemar (eds) *Vad krävs för att rädda demokratin?* Stockholm: Premiss förlag, pp 161–202.

Guston, David H. (1999). Stabilizing the Boundary between US Politics and Science: The Role of the Office of Technology Transfer as a Boundary Organization. *Social Studies of Science* 29(1): 87–112.

Guthman, Julie (2013). Fatuous Measures: The Artifactual Construction of the Obesity Epidemic. *Critical Public Health* 23(3): 263–273.

Haas, Peter M. (2007). Epistemic Communities. In: Bodanski, Daniel, Brunnée, Jutta & Hey, Ellen (eds) *The Oxford Handbook of International Environmental Law*. Oxford: Oxford University Press, pp 791–806.

Haas, Peter M. & Stevens, Casey (2011). Organized Science, Usable Knowledge and Multilateral Environmental Governance. In: Lidskog, Rolf & Sundqvist, Göran (eds) *Governing the Air: The Dynamics of Science, Policy, and Citizen Interaction*. Cambridge, MA: MIT Press, pp 125–161.

Habermas, Jürgen (1971). The Scientization of Politics and Public Opinion. In: Habermas, Jürgen, *Toward a Rational Society: Student Protest, Science and Politics*. Boston, MA: Beacon Press, pp 62–80.

Habermas, Jürgen (1984). *Theory of Communicative Action. Volume 1: Reason and the Rationalisation of Society*. Boston, MA: Beacon Press.

Habermas, Jürgen (1987). *Theory of Communicative Action. Volume II: The Critique of Functionalist Reason*. Cambridge: Polity Press.

Habermas, Jürgen (1996). *Between Facts and Norms: Contributions to a Discourse Theory of Law and Democracy*. Cambridge, MA: MIT Press.

Habermas, Jürgen (1997). Modernity: An Unfinished Project. In: Passerin d'Entrèves, Maurizio & Benhabib, Seyla (eds) *Habermas and the Unfinished Project of Modernity*. Cambridge, MA: MIT Press, pp 1–38.

Hammersley, Martyn (2022). Is 'Representation' a Folk Term? Some Thoughts on a Theme in Science Studies. *Philosophy of the Social Sciences* 52(3): 132–149.

Haraway, Donna J. (1996 [1988]). Situated Knowledges: The Science Question in Feminism and the Privilege of the Partial Perspective. In: Keller, Evelyn Fox & Longino, Helen E. (eds) *Feminism in Sciences*. Oxford: Oxford University Press, pp 249–263.

Haraway, Donna J. (2008). *When Species Meet*. Minneapolis, MN: University of Minnesota Press.

Hermansen, Erlend A.T., Lahn, Bård, Sundqvist, Göran & Øye, Eirik (2021). Post-Paris Policy Relevance: Lessons from the IPCC SR15 Process. *Climatic Change* 169(7): 1–18.

Hilgartner, Stephen (1990). The Dominant View of Popularization: Conceptual Problems, Political Uses. *Social Studies of Science* 20(3): 519–539.

Hilgartner, Stephen (2000). *Science on Stage: Expert Advice as Public Drama*. Stanford, CA: Stanford University Press.

Hilgartner, Stephen, Miller, Clark A. & Hagendijk, Rob (eds) (2015). *Science and Democracy: Making Knowledge and Making Power in the Biosciences and Beyond*. Abingdon: Routledge.

Horkheimer, Max & Adorno, Theodor W. (1944). *Dialectic of Enlightenment*. New York: Institute of Social Research.

Horst, Maja (2007). Public Expectations of Gene Therapy: Scientific Futures and their Performative Effects on Scientific Citizenship. *Science, Technology, & Human Values* 32(2): 150–171.

Horton, Richard (2020). *The COVID-19 Catastrophe: What's Gone Wrong and How to Stop It Happening Again*. Cambridge: Polity.

Hulme, Mike (2009). *Why We Disagree about Climate Change: Understanding Controversy, Inaction and Opportunity*. Cambridge: Cambridge University Press.

Illich, Ivan (1977). *Den Omänskliga Sjukvården: Det Medicinska Etablissemanget är ett Hot Mot Vår Hälsa och Inkräktar På Individens Rätt att Bestämma Över Sig Själv*. Stockholm: Aldus/Bonniers.

Irwin, Alan (1995). *Citizen Science: A Study of People, Expertise and Sustainable Development*. London: Routledge.

Irwin, Alan (2001). Constructing the Scientific Citizen: Science and Democracy in the Biosciences. *Public Understanding of Science* 10(1): 1–18.

Irwin Alan (2006). The Politics of Talk: Coming to Terms with the 'New' Scientific Governance. *Social Studies of Science* 36(2): 299–320.

Irwin, Alan & Wynne, Brian (eds) (1996). *Misunderstanding Science? The Public Reconstruction of Science and Technology*. Cambridge: Cambridge University Press.

Irwin, Alan & Michael, Mike (2003). *Science, Social Theory and Public Knowledge*. Maidenhead: Open University Press.

Isin, Engin F. (2009). Citizenship in Flux: The Figure of the Activist Citizen. *Subjectivities* 29(1): 367–388.

Isin, Engin F. and Turner, Bryan (eds) (2002). *Handbook of Citizenship Studies*. London: Sage.

Jasanoff, Sheila (1990). *The Fifth Branch: Science Advisers as Policymakers*. Cambridge, MA: Harvard University Press.

Jasanoff, Sheila (2003). Technologies of Humilities: Citizen Participation in Governing Science. *Minerva* 41(3): 223–244.

Jasanoff, Sheila (2004a). Ordering Knowledge, Ordering Society. In: Jasanoff, Sheila (ed) *States of Knowledge: The Co-Production of Science and Social Order*. London: Routledge, pp 13–45.

Jasanoff, Sheila (ed) (2004b). *States of Knowledge: The Co-Production of Science and Social Order*. London: Routledge.

Jasanoff, Sheila (2004c). The Idiom of Co-Production. In: Jasanoff, Sheila (ed) *States of Knowledge: The Co-Production of Science and Social Order*. London: Routledge, pp 1–12.

Jasanoff, Sheila (2005). *Designs of Nature: Science and Democracy in Europe and the United States*. Princeton, NJ: Princeton University Press.

Jasanoff, Sheila (2012). *Science and Public Reason*. Abingdon: Routledge.

Jasanoff, Shelia (2017). Science and Democracy. In: Felt, Ulrike, Fouché, Rayvon, Miller, Clark A. & Smith-Doerr, Laurel (eds) *The Handbook of Science and Technology Studies*. Cambridge, MA: MIT Press, pp 259–287.

Jasanoff, Sheila & Wynne, Brian (1998). Science and Decisionmaking. In: Rayner, Steve & Malone, Elizabeth L. (eds) *Human Choice and Climate Change, Vol 1: The Societal Framework*. Columbus, OH: Battelle Press, pp 1–87.

Kasperowksi, Dick & Hillman, Thomas (2018). The Epistemic Culture in an Online Citizen Science Project: Programs, Antiprograms and Epistemic Subjects. *Social Studies of Science* 48(4): 564–588.

Kasperowski, Dick & Hagen, Niclas (2019). Medborgarforskningens former. In: Soneryd, Linda & Sundqvist, Göran (eds) *Vetenskapligt Medborgarskap*. Lund: Studentlitteratur, pp 169–194.

Kera, Denisa (2012). Hackerspaces and DIYbio in Asia: Connecting Science and Community with Open Data, Kits and Protocols. *Journal of Peer Production* 1(2): 1–8.

Klintman, Mikael (2019). *Knowledge Resistance: How We Avoid Insights from Others*. Manchester: Manchester University Press.

Krueger, Tobias, Maynard, Carly, Carr, Gemma, Bruns, Antje, Mueller, Eva Nora & Lane, Stuart (2016). A Transdisciplinary Account of Water Research. *Wiley Interdisciplinary Reviews: Water* 3(3): 369–389.

Kuhn, Thomas S. (1962). *The Structure of Scientific Revolutions*. Chicago, IL: University of Chicago Press.

Landemore, Hélène (2020). *Open Democracy*. Princeton, NJ: Princeton University Press.

Latour, Bruno (1993). *We Have Never Been Modern*. Hemel Hempstead: Harvester Wheatsheaf.

Latour, Bruno (1996). The Flat-Earthers of Social Theory. In: Power, Michael (ed) *Accounting and Science: Natural Inquiry and Commercial Reason.* Cambridge: Cambridge University Press, pp xi–xvii.

Latour, Bruno (1998). From the World of Science to the World of Research? *Science* 280(5361): 208–209.

Latour, Bruno (1999a). When Things Strike Back – A Possible Contribution of 'Science Studies' to the Social Sciences. *British Journal of Sociology* 51(1): 105–123.

Latour, Bruno (1999b). *Pandora's Hope: Essays on the Reality of Science Studies.* Cambridge, MA: Harvard University Press.

Latour, Bruno (2004). *Politics of Nature: How to Bring the Sciences into Democracy.* Cambridge, MA: Harvard University Press.

Latour, Bruno & Schultz, Nikolaj (2022). *On the Emergence of an Ecological Class – a Memo.* Cambridge: Polity Press

Laurent, Brice (2017). *Democratic Experiments: Problematizing Nanotechnology and Democracy in Europe and the United States.* Cambridge, MA: MIT Press.

Leach, Melissa, Scoones, Ian & Wynne, Brian (2005). *Science and Citizens: Globalization and the Challenge of Engagement.* London: Zed Press.

Lezaun, Javier & Soneryd, Linda (2007). Consulting Citizens: Technologies of Elicitation and the Mobility of Publics. *Public Understanding of Science* 16(3): 279–297.

Liboiron, Max (2016). Redefining pollution and action: The matter of plastics. *Journal of Material Culture*, 21(1): 87–110.

Lidskog, Rolf & Sundqvist, Göran (2011). Science–Policy–Citizen Dynamics in International Environmental Governance. In: Lidskog, Rolf & Sundqvist, Göran (eds) *Governing the Air: The Dynamics of Science, Policy, and Citizen Interaction.* Cambridge, MA: MIT Press, pp 323–359.

Lidskog, Rolf & Sundqvist, Göran (2015). When Does Science Matter? International Relations Meets Science and Technology Studies. *Global Environmental Politics* 15(1): 1–20.

Lindblom, Charles E. (1959). The Science of 'Muddling Through'. *Public Administration Review* 19(2): 79–88.

Litfin, Karen T. (1995). *Ozone Discourses: Science and Politics in Global Environmental Cooperation.* New York, NY: Columbia University Press.

Longino, Helen E. (1990). *Science as Social Knowledge: Values and Objectivity in Scientific Inquiry.* Princeton, NJ: Princeton University Press.

Luhmann, Niklas (1989). *Ecological Communication.* Cambridge: Polity Press.

Luhmann, Niklas (1995). *Social Systems.* Stanford, CA: Stanford University Press.

Lundgren, Lars J. (1998). *Acid Rain on the Agenda: A Picture of a Chain of Events in Sweden 1966–1968.* Lund: Lund University Press.

Lynch, Michael (2017). STS, Symmetry and Post-Truth. *Social Studies of Science* 47(4): 593–599.

Lynch, Micheal, Hilgartner, Stephen & Berkowitz, Carin (2005). Voting Machinery, Counting and Public Proofs in the 2000 US Presidential Election. In: Latour, Bruno and Weibull, Peter (eds) *Making Things Public: Atmospheres of Democracy*. Cambridge, MA: MIT Press, pp 814–815.

Maasen, Sabine & Weingart, Peter (2005). What's New in Scientific Advice to Politics? In: Maasen, Sabine & Weingart, Peter (eds) *Democratization of Expertise? Exploring Novel Forms of Scientific Advice in Political Decision-Making*. Dordrecht: Springer, pp 1–19.

Marres, Noortje (2005) *No Issue, No Public: Democratic Deficits after the Displacement of Politics*. Doctoral dissertation, University of Amsterdam.

Marres, Noortje (2007). The Issues Deserve More Credit: Pragmatist Contributions to the Study of Public Involvement in Controversy. *Social Studies of Science* 37(5): 759–778.

Marres, Noortje (2012). *Material Participation: Technology, the Environment and Everyday Publics*. Basingstoke: Palgrave MacMillan

Martin, Ben R. & Etzkowitz, Henry (2000). The Origin and Evolution of the University Species. *VEST–Tidskrift för vetenskapsstudier* 13(3–4): 9–24.

Mastrandrea, Michael D., Heller, Nicole E., Root, Terry L. & Schneider, Stephen H. (2010). Bridging the Gap: Linking Climate-Impacts Research with Adaptation Planning and Management. *Climatic Change* 100(1): 87–101.

McGrath, Michael J. & Ni Scanaill, Cliodhna (2013). *Sensor Technologies: Healthcare, Wellness and Environmental Applications*. Berkeley, CA: Apress.

McNeil, Maureen & Haran, Joan (2013). Publics of Bioscience. *Science as Culture* 22(4): 433–451.

Merton, Robert K. (1945). The Role of the Intellectual in Public Bureaucracy. *Social Forces* 23(4): 405–415.

Merton, Robert K. (1973a [1942]). The Normative Structure of Science. In: Merton, Robert K., *The Sociology of Science: Theoretical and Empirical Investigations*. Chicago, IL: University of Chicago Press, pp 267–278.

Merton, Robert K. (1973b). *The Sociology of Science: Theoretical and Empirical Investigations*. Chicago IL: University of Chicago Press.

Metzger, Jonathan, Soneryd, Linda & Linke, Sebastian (2017). The Legitimization of Concern: A Flexible Framework for Investigating the Enactment of Stakeholders in Environmental Planning and Governance Processes. *Environment and Planning A* 49(11): 2517–2535.

Miller, Clark A. (2004). Climate Science and the Making of a Global Political Order. In: Jasanoff, Sheila (ed) *States of Knowledge: The Co-Production of Science and Social Order*. London: Routledge, pp 46–66.

Mirowski, Philip (2018). The Future(s) of Open Science. *Social Studies of Science* 48(2): 171–203.

Moore, Alfred (2010). Beyond Participation: Opening Up Political Theory in STS. *Social Studies of Science* 40(5): 793–799.

Nacka Tingsrätt – Mark- och miljödomstolen (2018). *Mark- och miljödomstolens yttrande 2018-01-23. Mål nr M 1333-11. Aktbilaga 842.* Saken: Tillstånd enligt miljöbalken till anläggningar i ett sammanhängande system för slutförvaring av använt kärnbränsle och kärnavfall; nu fråga om yttrande till regeringen.

Naturvårdsverket (2009). *Allmänheten och klimatförändringen 2009. Allmänhetens kunskap om och attityd till klimatförändringen, med fokus på egna åtgärder, konsumtionsbeteenden och företagens ansvar.* Report 6311. Stockholm: Naturvårdsverket.

Nelkin, Dorothy (1975). The Political Impact of Technical Expertise. *Social Studies of Science* 5(1): 35–54.

Nelkin, Dorothy (ed) (1984). *Controversy: Politics of Technical Decisions.* Beverly Hills, CA: Sage.

Nelkin, Dorothy (1995). Science Controversies: The Dynamics of Public Disputes in the United States. In: Jasanoff, Sheila, Markle, Gerald, Peterson, James & Pinch, Trevor (eds) *Handbook of Science and Technology Studies.* Thousand Oaks, CA: Sage, pp 444–456.

Nowotny, Helga, Scott, Peter & Gibbons, Michael (2001). *Re-Thinking Science: Knowledge and the Public in an Age of Uncertainty.* Cambridge: Polity Press.

Odén, Svante (1967). Nederbördens försurning. *Dagens Nyheter*, 24 October.

Ottinger, Gwen (2010). Buckets of Resistance: Standards and the Effectiveness of Citizen Science. *Science, Technology and Human Values* 35(2): 244–270.

Pallett, Helen & Chilvers, Jason (2022). STS and Democracy Co-produced? The Making of Public Dialogue as a Technology of Participation. In: Birkbak, Andreas & Papazu, Irina (eds) *Democratic Situations.* Manchester: Mattering Press, pp 118–140.

Peterson, Alan & Lupton, Deborah (1996). *The New Public Health: Health and Self in the Age of Risk.* Thousand Oaks, CA: Sage.

Petersson, Jesper (2019). Sensorer, luftkvalitet och vetenskapligt medborgarskap. In: Soneryd, Linda & Sundqvist, Göran (eds) *Vetenskapligt Medborgarskap.* Lund: Studentlitteratur, pp 195–217.

Petryna, Adriana (2002). *Life Exposed: Biological Citizens After Chernobyl.* Princeton, NJ: Princeton University Press

Pfotenhauer, Sebastian, Laurent, Brice, Papageorgiou, Kyriaki & Stilgoe, Jack (2022). The Politics of Scaling. *Social Studies of Science* 52(1): 3–34.

Pinch, Trevor (2001). Scientific Controversies. In: Smelser, Neil J. & Baltes, Paul B. (eds) *International Encyclopedia of the Social and Behavioral Sciences.* Amsterdam: Elsevier, pp 13719–13724.

Plato (2007). *The Republic.* London: Penguin Classics.

Polanyi, Michael (1962). The Republic of Science: Its Political and Economic Theory. *Minerva* 1(1): 54–73.

Polanyi, Michael (1966). *The Tacit Dimension.* Chicago, IL: University of Chicago Press.

Prainsack, Barbara (2014). Understanding Participation: The 'Citizen Science' of Genetics. In: Prainsack, Barbara, Schicktanz, Silke & Werner-Felmayer, Gabriele (eds) *Genetics as Social Practice*. Farnham: Ashgate, pp 147–164.

Price, Don K. (1965). *The Scientific Estate*. Cambridge, MA: Harvard University Press.

Prützer, Madeleine & Soneryd, Linda (2016). *Samverkan och Deltagande i Vattenråd och Vattenförvaltning*. Gothenburg: Havs- och vattenmyndigheten.

Rabinow, Paul (1992). Artificiality and Enlightenment: From Sociobiology to Biosociality. In: Crary, Jonathan & Kwinter, Sanford (eds) *Incorporations*. New York: Zone Books.

Rancières, Jacques (2009). *Hatred of Democracy*. London: Verso.

Ravetz, Jerry (2022). Science – Post-Normal Perspectives. *Futures* 140: 102958.

Reynolds, Larry (2013). The Contested Publics of the UK GM Controversy: A Tale of Entanglement and Purification. *Science as Culture* 22(4): 452–475.

Rosanvallon, Pierre (2010). *Counter-Democracy: Politics in an Age of Distrust*. Cambridge: Cambridge University Press

Rose, Dale & Blume, Stuart (2003). Citizens as Users of Technology: An Exploratory Study of Vaccines and Vaccination. In: Oudshoorn, Nelly & Pinch, Trevor (eds) *How Users Matter: The Co-Construction of Users and Technologies*. Cambridge, MA: MIT Press, pp 103–131.

Rose, Nikolas (1999). *Powers of Freedom: Reframing Political Thought*. Cambridge: Cambridge University Press.

Rose, Nikolas & Novas, Carlos (2004). Biological Citizenship. In: Ong, Aihwa & Collier, Stephen J. (eds) *Global Assemblages: Technology, Politics, and Ethics as Anthropological Problems*. Oxford: Blackwell Publishing, pp 439–463.

Rudig, Wolfgang & Flam, Helena (eds) (1994). *States and Anti-Nuclear Movements*. Edinburgh: Edinburgh University Press.

Ryghaug, Marianne, Skjølsvold, Tomas Moe & Heidenreich, Sara (2018). Creating Energy Citizenship through Material Participation. *Social Studies of Science* 48(2): 283–303.

Salomon, Jean-Jacques (2000). Science, Technology and Democracy. *Minerva* 38(1): 33–51.

Salter, Liora (1988). *Mandated Science: Science and Scientists in the Making of Standards*. Dordrecht: Kluwer Academic Publishers.

SAPEA (Science Advice for Policy by European Academies) (2019). *Making Sense of Science for Policy under Conditions of Complexity and Uncertainty*. Evidence Review Report No 6. Berlin: SAPEA.

Sarewitz, Daniel (2004). How Science Makes Environmental Controversies Worse. *Environmental Science and Policy* 7(5): 385–403.

Sarewitz, Daniel & Pielke, Roger A., Jr (2007). The Neglected Heart of Science Policy: Reconciling Supply of and Demand for Science. *Environmental Science and Policy* 10(1): 5–16.

Sassower, Raphael (2014). *The Price of Public Intellectuals*. Basingstoke: Palgrave Macmillan.

Schnaiberg, Allan (1977). Obstacles to Environmental Research by Scientists and Technologists: A Social Structural Analysis. *Social Problems* 24(5): 500–520.

Seifert, Franz (2006). Local Steps in an International Career: A Danish-Style Consensus Conference in Austria. *Public Understanding of Science* 15(1): 73–88.

Shapin, Steven (1996). *The Scientific Revolution*. Chicago, IL: University of Chicago Press.

Shapin, Steven (2007). Expertise, Common Sense, and the Atkins Diet. In: Porter, Jean & Phillips, Peter (eds) *Public Science in Liberal Democracy*. Toronto: University of Toronto Press, pp 174–193.

Shapin, Steven (2010). *Never Pure: Historical Studies of Science as if It Was Produced by People with Bodies, Situated in Time, Space, Culture, and Society, and Struggling for Credibility and Authority*. Baltimore, MD: Johns Hopkins University Press.

Shapin, Steven & Schaffer, Simon (1985). *Leviathan and the Air-Pump*. Princeton, NJ: Princeton University Press.

Sismondo, Sergio (2010). *An Introduction to Science and Technology Studies* (2nd edn). Oxford: Wiley-Blackwell.

Sismondo, Sergio (2017). Post-truth? *Social Studies of Science* 47(1): 3–6.

Smart, Palie, Holmes, Sara, Lettice, Fiona, Pitts, Frederick Harry, Zwiegelaar, Jeremy Basil, Schwartz, Gregory & Evans, Stephen (2019). Open Science and Open Innovation in a Socio-Political Context: Knowledge Production for Societal Impact in an Age of Post-Truth Populism. *R&D Management* 49(3): 279–297.

Söderberg, Johan (2019). Olaglig innovation bland droganvändare. In: Soneryd, Linda & Sundqvist, Göran (eds) *Vetenskapligt medborgarskap*. Lund: Studentlitteratur, pp 219–242.

Soneryd, Linda (2007). Deliberations over the Unknown, the Unsensed and the Unsayable? Public Protests and the 3G Development in Sweden. *Science, Technology and Human Values* 32(3): 287–314.

Soneryd, Linda (2016). Technologies of Participation and the Making of Technologized Futures. In: Chilvers, Jason & Kearnes, Matthew (eds) *Remaking Participation: Science, Environment and Emergent Publics*. London: Routledge, pp 144–161.

Soneryd, Linda & Uggla, Ylva (2015). Green Governmentality and Responsibilization: New Forms of Governance and Responses to 'Consumer Responsibility'. *Environmental Politics* 24(6): 913–931.

Soneryd, Linda & Amelung, Nina (2016). Translating Participation: Scenario Workshops and Citizens' Juries across Situations and Contexts. In: Voß, Jan-Peter & Freeman, Richard (eds) *Knowing Governance: The Epistemic Construction of Political Order*. Basingstoke: Palgrave Macmillan, pp 155–174.

Soneryd, Linda & Sundqvist, Göran (eds) (2019). *Vetenskapligt medborgarskap*. Lund: Studentlitteratur.

Soneryd, Linda & Sundqvist, Göran (2022). Leaks and Overflows: Two Contrasting Cases of Hybrid Participation in Environmental Governance. In: Birkbak, Andreas & Papazu, Irina (eds) *Democratic Situations*. Manchester: Mattering Press, pp 101–117.

Stehr, Nico & Grundmann, Reiner (2012). How Does Knowledge Relate to Political Action? *Innovation: The European Journal of Social Science Research* 25(1): 29–44.

Stevenson, Harley & Dryzek, John S. (2014). *Democratizing Global Climate Governance*. Cambridge: Cambridge University Press.

Stichweh, Rudolf (2022). *Functional Differentiation of Society*. Bielefeld: Transcript Verlag.

Stilgoe, Jack (2007). *Nanodialogues: Experiments in Public Engagement with Science*. London: Demos.

Stirling, Andy (2008). 'Opening Up' and 'Closing Down': Power, Participation, and Pluralism in the Social Appraisal of Technology. *Science, Technology, & Human Values* 33(2): 262–294.

Strathern, Marilyn (2006). Bullet-Proofing: A Tale from the United Kingdom. In: Riles, Annelise (ed) *Documents: Artefacts of Modern Knowledge*. Ann Arbor, MI: Michigan University Press, pp 181–205.

Sundqvist, Göran (2000). The Environmental Expert. In: Lundgren, Lars J. (ed) *Knowing and Doing: On Knowledge and Action in Environmental Protection*. Stockholm: Swedish Environmental Protection Agency, pp 51–73.

Sundqvist, Göran (2002). *The Bedrock of Opinion: Science, Technology and Society in the Siting of High-Level Nuclear Waste*. Dordrecht: Kluwer Academic Publishers.

Sundqvist, Göran (2012). Governing Unruly Technology: Swedish Politicians and Nuclear Power. In: Larsson, Bengt, Letell, Martin & Thörn, Håkan (eds) *Transformations of the Swedish Welfare State: From Social Engineering to Governance?* Basingstoke: Palgrave Macmillan, pp 56–70.

Sundqvist, Göran (2014). 'Heating Up' or 'Cooling Down'? Social Scientists Analysing and Performing Environmental Conflicts. *Environment & Planning A* 46(9): 2065–2079.

Sundqvist, Göran (2019). Speaking Truth to Power: Science and Policy. In: Ritzer, Georg and Rojek, Chris (eds) *The Blackwell Encyclopedia of Sociology*. Chicester: John Wiley & Sons.

Sundqvist, Göran (2021). *Vem Bryr Sig? Om Klimatforskning och Klimatpolitik*. Gothenburg: Daidalos.

Sundqvist, Göran & Elam, Mark (2009). Sociologin, hybriderna och den sociala verkligheten. Exemplet kärnavfall. *Sociologisk Forskning* 46(2): 4–25.

Sundqvist, Göran, Bohlin, Ingemar, Hermansen, Erlend A.T. & Yearley, Steven (2015). Formalization and Separation: A Systematic Basis for Interpreting Approaches to Summarizing Science for Climate Policy. *Social Studies of Science* 45(3): 416–440.

Sundqvist, Göran, Gasper, Des, St Clair, Asunción L., Hermansen, Erlend A.T., Yearley, Steven, Tvedten, Irene Ø. & Wynne, Brian (2018). One World or Two? Science-Policy Interactions in the Climate Field. *Critical Policy Studies* 12(4): 448–468.

Swedish Government (2019). *Regeringens proposition 2019/20:65. En samlad politik för klimatet – klimatpolitisk handlingsplan.*

Tait, Joyce (2009). *Upstream Engagement and the Governance of Science: The Shadow of the Genetically Modified Crops Experience in Europe. EMBO Reports* 10: S18–S22.

The Guardian (2019). 'Listen to the Scientists': Greta Thunberg Urges Congress to Take Action, 19 September.

The Washington Post (2017). President Trump Has Made 1,318 False or Misleading Claims over 263 Days, 10 October.

Trevino, A. Javier & Staubmann, Helmut (2022). *The Routledge International Handbook of Talcott Parsons Studies.* Abingdon: Routledge.

Uggla, Ylva (2008). Strategies to Create Risk Awareness and Legitimacy: The Swedish Climate Campaign. *Journal of Risk Research* 11(6): 719–734.

Ureta, Sebastian, Flores, Patricio & Soneryd, Linda (2020). Victimization Devices: Exploring Challenges Facing Litigation-Based Transnational Environmental Justice, *Social and Legal Studies* 29(2): 161–182.

Venturini, Tommaso and Munk, Anders Kristian (2021). *Controversy Mapping: A Field Guide.* Cambridge: Polity.

Vihersalo, Mirja (2017). Climate Citizenship in the European Union: Environmental Citizenship as an Analytical Concept, *Environmental Politics* 26(2): 343–360.

Walzer, Michael (1983). *Spheres of Justice: A Defense of Pluralism and Equlity.* New York: Basic Books.

Weart, Spencer R. (2008). *The Discovery of Global Warming.* Cambridge, MA: Harvard University Press.

Weber, Max (1983). *Ekonomi och samhälle. Förståelsesociologins grunder. Sociologiska begrepp och definitioner. Ekonomi, samhällsordning och Grupper.* Lund: Argos.

Weber, Max (2001 [1930]). *The Protestant Ethic and the Spirit of Capitalism.* London: Routledge.

Weber, Max (2004). *The Vocation Lectures: Science as a Vocation – Politics as a Vocation.* Cambridge: Hacket Publishing.

Weingart, Peter (1999). Scientific Expertise and Political Accountability: Paradoxes of Science in Politics. *Science and Public Policy* 26(3): 151–161.

Welsh, Ian & Wynne, Brian (2013). Science, Scientism and Imaginaries of Publics in the UK: Passive Objects, Incipient Threats. *Science as Culture*, 22(4): 540–566.

Wilsdon, James & Willis, Rebecca (2004). *See-Through Science: Why Public Engagement Needs to Move Upstream*. London: Demos.

Woolgar, Steve (2014). Struggles with Representation: Could It Be Otherwise? In: Coopman, Catelijne, Vertesi, Janet, Lynch, Michael & Woolgar, Steve (eds) *Representation in Scientific Practice Revisited*. Cambridge, MA: MIT Press, pp 329–332.

World Health Organization (2022) *Ambient (Outdoor) Air Pollution*. Fact sheet, 19 December 2022. Available at: https://www.who.int/news-room/fact-sheets/detail/ambient-(outdoor)-air-quality-and-health

Wynne, Brian (1987). *Risk Management and Hazardous Waste: Implementation and the Dialectics of Credibility*. Berlin: Springer Verlag.

Wynne, Brian (1991). Knowledges in Context. *Science, Technology and Human Values* 16(2): 111–121.

Wynne, Brian (1993). Public Uptake of Science: A Case for Institutional Reflexivity. *Public Understanding of Science* 2(4): 321–337.

Wynne, Brian (1995). Public Understanding of Science. In: Jasanoff, Sheila, Markle, Gerald, Peterson, James & Pinch, Trevor (eds) *Handbook of Science and Technology Studies*. Thousand Oaks, CA: Sage, pp 361–388.

Wynne, Brian (1996). May the Sheep Safely Graze? A Reflexive View of the Expert-Lay Knowledge Divide. In: Lash, Scott, Szerszynski, Bronislaw & Wynne, Brian (eds) *Risk, Environment and Modernity: Towards a New Ecology*. London: Sage, pp 44–83.

Wynne, Brian (2001). Creating Public Alienation: Expert Cultures of Risk and Ethics on GMOs. *Science as Culture* 10(4): 445–481.

Wynne, Brian (2007). Public Participation in Science and Technology: Performing and Obscuring a Political-Conceptual Category Mistake. *East Asian Science, Technology and Society* 1(1): 99–110.

Wynne, Brian (2016). Ghosts of the Machine: Publics, Meanings and Social Science in a Time of Expert Dogma and Denials. In: Chilvers, Jason & Kearnes, Matthew (eds) *Remaking Participation: Science, Environment and Emergent Publics*. London: Routledge, pp 99–120.

Yamin, Farhana & Depledge, Joanna (2004). *The International Climate Change Regime: A Guide to Rules, Institutions and Procedures*. Cambridge: Cambridge University Press.

Ziman, John (1994). *Prometheus Bound: Science in a Dynamic Steady State*. Cambridge: Cambridge University Press.

Index

Printed and bound by CPI Group (UK) Ltd, Croydon, CR0 4YY
27/10/2024

14580559-0005